"MIND, HEART, AND VISION"

Professional Engineering in Canada 1887 to 1987

This late nineteenth century photograph
of Civil Engineer James McDougall and the
instruments of his calling is a study in
professional pride. By 1887, when they
formed their first professional society,
Canadian engineers had already created an
enviable legacy of achievement and looked
forward to even greater accomplishments.
(PAC PA-147706)

"MIND, HEART, AND VISION"

Professional Engineering in Canada 1887 to 1987

NORMAN R. BALL

National Museum of Science and Technology,
National Museums of Canada,
in cooperation with
the Engineering Centennial Board

Canadian Cataloguing in Publication Data

Ball, Norman R., 1944-
  "MIND, HEART, AND VISION" Professional Engineering in Canada 1887 to 1987

Issued also in French under title: « VISION, COEUR ET RAISON »
  L'ingénierie au Canada de 1887 à 1987
ISBN 0-660-12000-3

1. Engineering – Canada – History. 2. Engineers – Canada. I. National Museum of Science and Technology (Canada). II. Engineering Centennial Board (Canada). III. Title.

TA26 B34 1987   620′.00971   C87-099107-8

Contents

Foreword

Few countries have demanded more of their engineers than Canada. This valiant group of people has made it possible to communicate and travel across our vast country. They have developed techniques for extracting natural resources and for fashioning these resources into manufactured goods. They have supplied us with such plentiful sources of power that we now sell electricity abroad. They have housed our families and created places for us to work in this often inhospitable land. And they have reached out from Canada to help other parts of the world develop as we have.

The products of their work have been useful and often extremely beautiful. But, like the air we breathe, engineering accomplishments become part of the background of our everyday lives. Too often, we take engineers for granted and fail to recognize their creativeness.

I consider it a great honour that the National Museum of Science and Technology was chosen to participate in the Centennial Celebration of Engineering in Canada and, in particular, to bring to all Canadians this history of Canadian engineering's first hundred years as an organized profession. I am particularly proud of Dr. Norman Ball, Chief of Research, and our publications staff for their dedicated efforts in making this project a success.

The Museum's job is to bring to all Canadians and people in other countries the often untold story of accomplishments in technology and its supporting sciences. We can best achieve this goal by working closely with industries, universities, and federal, provincial, and regional governments. The cooperation and support of the Engineering Centennial Board and of the engineering community in Canada have contributed a great deal to the realization of this project.

I am certain that all of you who read of the success of our Canadian engineers join me in a salute to their marvelous achievements. This first century is but a taste of what will be accomplished in the next one.

J.Wm. McGowan
Director
National Museum of Science and Technology

Financial assistance for centennial activities from the following corporations is gratefully acknowledged:

Noranda Inc.

The Canam Manac Group Inc.

Acres International Limited

Dofasco Inc.

Teleglobe Canada

Canadian Pacific

Imperial Oil

and most particularly the support of
Energy, Mines and Resources Canada

Mind, Heart, and Vision

A hundred years ago a handful of individuals created the first association of engineers in our country: the Canadian Society of Civil Engineers. It marked the beginning of a tradition and the first attempt to gain recognition for our profession.

The following pages recapture the saga of our evolution since that time. They shed a new light on the important role played by engineers in Canadian history, a role seldom acknowledged in the history books. By publishing a book dedicated entirely to engineers and their achievements over the last century, we seek to remedy this oversight. Most of the information skilfully gathered and brought to life by the author, Dr. Norman Ball, to whom we are most grateful, has never been publicized. It reveals how the mind, heart, and vision of our engineers succeeded in shaping this country through innovative and resourceful endurance.

We are confident that this first effort will not be the last, that it will bring a new understanding of our profession, an awareness of our contribution and involvement in all facets of Canadian life, and that it will inspire other research and publications on our great profession.

On behalf of the Engineering Centennial Board, I am very honoured to present to you this tribute to our peers. The future of engineering is intimately linked to that of Canadian society. In this respect, the theme of our centennial, "Mind, Heart, and Vision – Engineering in Canada: The Next Hundred Years," reflects our commitment to the community.

Bernard Lamarre
Chairman
Engineering Centennial Board

Acknowledgements

Were it not for the vision and support of Bernard Lamarre, Chairman of the Engineering Centennial Board, Inc., there would be no book. I am deeply indebted to Mr. Lamarre for the great pleasure of working with him and his colleagues. Dr. J. William McGowan is also to be thanked for making the centennial of engineering as an organized profession in Canada an occasion to commit the National Museum of Science and Technology to a new level of dedication to historical research and publication.

For their research input at various stages, I am grateful to Ruth Walker, Ken Wallace, Tom Roach, and John Collins. I am unable to list all who shared their knowledge as research progressed, but would like to acknowledge the special efforts made by Jane Hainsworth, Mary Hall, Heather Hatch, Sylvia Macenko, Dawn Monroe, Robert Enns, Larry McNally, David Neufeld, and Robert Passfield.

Unlike Euclidean geometry with its numerous theorems, propositions, and corollaries, writing books to a deadline has only one unalterable rule: things go wrong. There was no shortage of surprises and a group effort kept everything moving. Dr. Léopold Nadeau took everything in stride. Wendy McPeake, project manager, kept us pointing in the same direction and on time. John Collins's photo research work and Norman Dahl's early photo editing contributed immensely to the final shape of the book. The book designer, Eiko Emori, transformed pages of typed text and binders of photographs, with a skill akin to magic, into a visual treat.

To the many engineers and companies who supplied achievements, information, or both, I give my sincere thanks and the promise that there will be more books to do greater justice to them; they have done more than one can capture between the covers of one book.

Both Mr. Collins and I would like to thank the companies and institutions that were so helpful in our search for illustrations. It was a pleasure to find so much enthusiasm about our work and a source of regret that we could not use all that we were offered. The source of each illustration is indicated at the end of its caption.

My family has been most important during both ups and downs. As research files spilled out into the halls and I retreated deeper into the study, my wife Pat was amazingly patient, understanding, and, most important, there. Our son Alexander contributed his good humour, interest in how things work, and companionship on the bicycle rides that provided relaxation when writing did not. Our daughter Heather managed to put everything in perspective as only a child can do when she asked: "Does this book come with a party? Remember how we went for a holiday and got to go to a grown-up party when you did the hospital book?" Yes Heather, there will be party.

Norman R. Ball

Introduction

This book is written on the occasion of an important centennial – the founding of Canada's first professional engineering society, the Canadian Society of Civil Engineers, in 1887. The formation of the Society marked a significant step in the growth of Canadian engineering. But even before 1887, the country's immense distances, harsh climate and topography, and frontier economic conditions had begun to mould a distinctive and valuable engineering tradition in Canada.

The development of this tradition from the early nineteenth century to the present day and the way in which engineering forms an integral part of Canadian history are the subjects of this book. It is an exciting and remarkable story. The creation of a modern industrial state on the inhospitable northern half of North America is largely a testimony to the excellence of Canadian engineering.

Yet, as with many of the extraordinary achievements of Canadians, Canada's engineering heroes and their accomplishments remain largely unsung. Canadians are well-known for their reluctance to celebrate or even recognize excellence in their own country. In fact, despite its evident importance, most Canadian historical studies have tended to ignore engineering.

Both Canadian engineering achievements and their relative neglect make it appropriate to publish an historical introduction to the topic on the hundredth anniversary of founding of the first professional society. The aim of this book is not to chronicle every major engineering development in Canada – this would be well beyond the scope of such a short text. Rather, it is an introduction to engineering excellence and the essential role engineering has played in Canadian history.

One can only hope that this book will help Canadian engineering gain the recognition it so richly deserves during its second century as an organized profession.

1

Bold Beginnings: The Years before 1887

The formation in 1887 of Canada's first national professional engineering society, the Canadian Society of Civil Engineers, was an important benchmark in Canadian engineering history. But it did not signal the beginning of great engineering achievement in Canada. On the contrary. Canadian engineers, decades earlier, in surmounting the unprecedented challenges posed by the country's geography, climate, and history, had already begun to create, through the construction of their remarkable dams, canals, and railways, a unique and valuable tradition which continues to influence engineering to this day.

When Canadians look back in history to great and memorable feats of engineering in this country, they are apt to think solely in terms of railroads in general and the Canadian Pacific Railway in particular. Although the building of the transcontinental railroad in the late nineteenth century was of major historical importance, and has been justly celebrated as such, it has tended to minimize the significance of the outstanding groundbreaking work of earlier generations of engineers.

The inadequacy of the conventional view was underlined almost 30 years ago by the eminent engineer and historian, Dr. Robert Legget. Speaking to the Newcomen Society in London, England, Dr. Legget said, "The completion of the Canadian Pacific Railway through the western mountains in 1885 has been so well publicized that it is natural that many should think of this great feat as the first major engineering undertaking of the Dominion."[1] To provide a more accurate and comprehensive understanding of Canadian engineering history, Dr. Legget went on to speak about earlier railroads, canals, and military works. His primary example was the Rideau Canal and, in particular, the Jones Falls Dam, one of the 52 dams on the canal. Still in use today, the Rideau Canal remains a particularly striking example of early engineering achievement in Canada.

The Rideau Canal: An Engineering Triumph
Much of Canada's early engineering history revolves around the creation of transportation and communications systems in the face of staggering challenges. The Rideau Canal was conceived as part of a defensive military network linking the strategic city of Kingston at the foot of Lake Ontario, via Bytown on the Ottawa River, to Montreal. This provided a safe water route, away from U.S. shores, to the St. Lawrence and ultimately to the Atlantic. When it opened in 1832, the Rideau Canal was one of the world's greatest feats of engineering and construction.

The Jones Falls Dam near Kingston, cited by Dr. Legget, is an excellent place to start an examination of this achievement. The Jones Falls Dam was the first true arched masonry dam in North America, the tallest dam in the western hemisphere, and the largest of its kind in the world. Located far from any settlement in the middle of virgin forest and malaria-ridden swamp, the dam was a noteworthy accomplishment in its own right and it marked a turning point in Canadian engineering history.

The Jones Falls Dam was built of massive sandstone blocks quarried eight kilometres away and drawn by oxen and scows to the site. The dam is a graceful, imposing arch stretching 107 metres along its crest and rising to a height of 19 metres. Many arched dams are built up from the base of narrow gorges shaped like an inverted triangle. Because it was built through a wider cut, the Jones Falls Dam was significantly more difficult to construct.

The Jones Falls Dam demonstrates one of the longlasting trends in Canadian engineering. Not only was the dam the largest of its kind in the world, but the locks themselves were also the highest. Other locks would take larger ships but none in steps as high as the four locks at Jones Falls. Building the biggest, along with

Originally constructed for defensive purposes, the Rideau Canal took advantage of the Rideau River and lake system, connecting unnavigable sections with canals and locks. This map of its route, connecting Bytown in the lower right and Kingston in the lower left, was made from an 1830 drawing by Colonel John By, the Royal Engineer in charge of the project. Colonel By's signature can be seen on the lower right. (Public Archives Canada NMC-0043041)

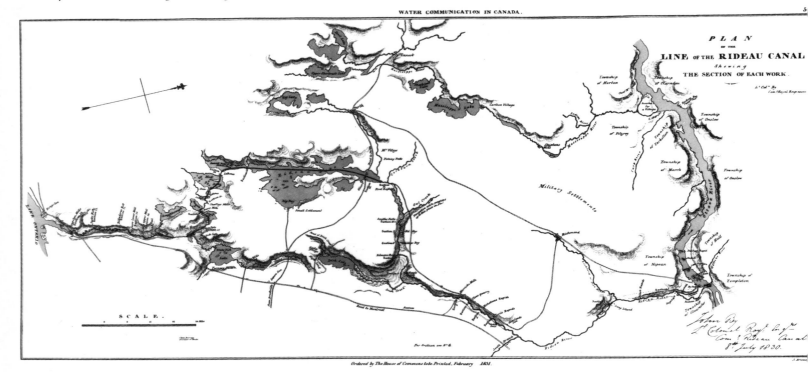

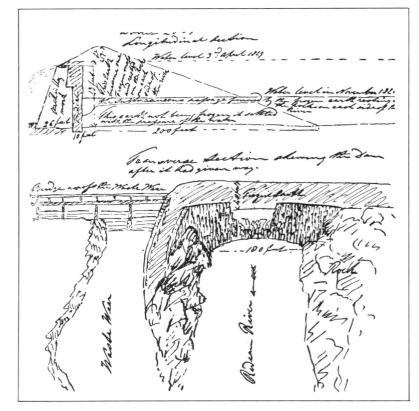

Three failures of the Hog's Back Dam on the Rideau River taught Colonel By and his colleagues that engineering in Canada's harsh wilderness environment required a departure from traditional European practices. This sketch is taken from a letter written by Colonel By describing his numerous problems at the site. Originally conceived as a stone-arch dam, Hog's Back finally emerged as a more readily constructed stone-filled, timber-crib dam. (Public Archives Canada)

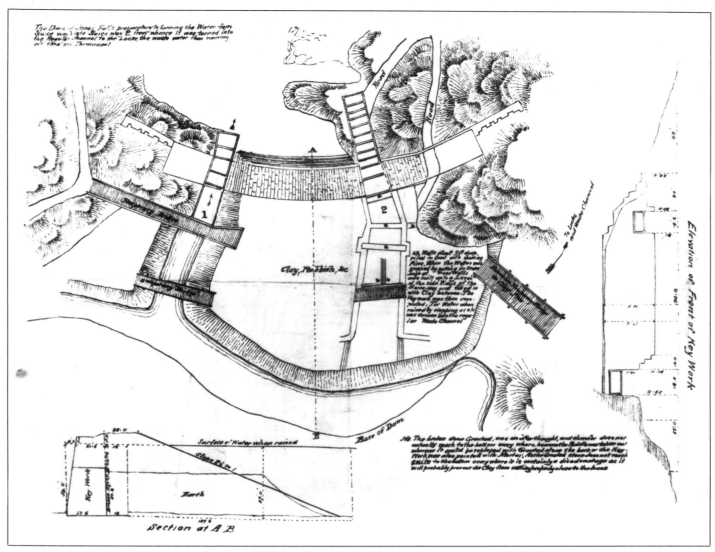

The Jones Falls Dam on the Rideau Canal was built following an engineering tradition brought from Europe which emphasized permanence, finish, and attention to detail. The graceful curvature of the dam can be seen in the photograph above and in the upper portion of the engineering drawing at the right. The drawing, which dates from the time of construction, contains descriptions of some of the engineering techniques employed on the dam. The masonry key-work, for example, was sunk 2.5 metres into the riverbed to secure a solid rock foundation and was built up with blocks of dressed stone 1.9 m by 46 cm. These were laid on end in tightly joined vertical courses. (Parks Canada, N.A. Patterson Collection, R4-020-F-0055, R4-020-F-0056)

the engineering, construction, and management skills to work on a hitherto unknown scale, was going to become a Canadian engineering trademark.

The dam's singular significance is, however, that it represents the end of a transplanted European engineering tradition and the beginning of an indigenous Canadian one. Colonel John By, the Royal Engineer in charge of the design and construction of the Rideau Canal, often spoke of the canal as being built "to stand the test of time." "Built to last" was an important part of the European engineering tradition which had developed in densely populated and long-settled areas close to existing centres of trade, manufacturing, and skills. It implied the use of materials to give both the appearance and the reality of permanence. It required maximum financial resources up front and minimum long-term maintenance costs.

The Jones Falls Dam embodied the best of this European tradition. But as the job progressed, it became an accountant's nightmare. Traditional modes of planning and execution were not enough to compensate for Canada's fierce climate and the fact that the country lacked the economic, physical, and human resources taken for granted in much of Europe.

Despite tremendous costs, the Jones Falls Dam was completed in accordance with the older engineering traditions. But when Colonel By tried to use these same traditions to build another dam on the Rideau Canal – the Hog's Back Dam in Ottawa – combined environmental forces overwhelmed him and left no doubt as to the inadequacy of transplanted European engineering.

Construction at Hog's Back, where the Rideau River narrows sharply and drops precipitously, started in July 1827. Loose earth and stone were dumped into the river and gradually a masonry arch of closely fitted cut stone began to take shape. In February 1828, a sudden flood tore out much of the fill and part of the

11.3-metre-high masonry structure. As an emergency measure, the 15th Company of Royal Sappers and Miners rapidly plugged the gap with a rock-filled timber crib which was then destroyed by a flood in March of 1828. Colonel By stuck with his design. Work resumed on the stone key-work of his arched masonry dam, and in the following year the customarily violent spring floods again destroyed his work.

Eventually he succeeded. In the spring of 1831, a dam 15 metres high stretched 76 metres across the river. But it bore no resemblance to By's original design. The Hog's Back Dam was characteristically Canadian.

Colonel By's fundamental problem was that, with the construction equipment then available, conventional stone arches simply took too long to bring to a point where they could withstand the violent spring floods of most Canadian rivers. In the spring, the waters of Hog's Back usually rose 4.3 to 4.9 metres above their normal level; British engineers were accustomed to floods of about 0.6 metres. Furthermore, although good stonemasons and readily available stone were in short supply in Canada, skilled axemen and timber were obtainable in abundance.

The successful Hog's Back Dam took advantage of the available resources. It consisted of a series of quickly built timber cribs filled with stone rubble. To those accustomed to the smooth regularity and well-crafted appearance of fine stonework, the rubble-filled timber crib dam appeared unsightly and unworkmanlike. To those who understood the resources and conditions, it was an excellent piece of engineering.

The Hog's Back Dam demonstrated an important lesson for Canadian engineers: good engineering design and technology are adapted to the circumstances. By 1830, the pressure of soaring costs led to the use of earth fill for dams and timber floors for locks, instead of the more expensive stone. Canadian engineering was becoming attuned to the requirements of an emerging nation for quick, functional construction at reasonable costs.

John MacTaggart, Visionary

John MacTaggart, Clerk of the Works and Colonel John By's second in command, appears to have been largely responsible for creating the radical new vision of engineering that began to take shape during the construction of the Rideau Canal.

MacTaggart was a curious character. It is rumoured that he drank too much. He left before the job was finished, returned to Scotland in broken health, and died in 1830. Yet, after he finished work on the Rideau Canal, MacTaggart published a two-volume general work about Canada based on his experiences.[2] The book provides an early and prophetic insight into engineering in Canada.

MacTaggart was intensely aware of Canada's need for the imaginative use of technology. His writing contains a wealth of suggestions for solutions to specific economic and engineering problems, with applications on the larger scale as well. Sovereignty was one of his primary concerns. Canada's closest neighbour had chosen revolution as the means to end its colonial status and was pursuing a markedly expansionist course. In MacTaggart's mind, technology and sovereignty were intimately connected: the effective use of technology was essential for defining the country and for keeping it intact. MacTaggart suggested, for example, that Canada establish ''rights to disputed lands'' by building a network of telegraph stations stretching from Cape Breton Island to the Pacific Ocean.

MacTaggart saw that Canadian engineering must pursue knowledge relevant to its own requirements rather than view the country through the eyes, habits, and traditions of others. His vision of Canada was twofold.

First, he said, get to know the country, its resources, and assets. Second, use this knowledge imaginatively. Over 150 years ago, he perceived this path as the means of transforming Canada from a supplier of raw materials to an industrial nation. A few examples will clarify how he viewed the role of engineering in this grand design.

On the voyage across the Atlantic to Canada, MacTaggart passed by Newfoundland and the Grand Banks. Impressed by the area's abundant maritime resources, he commented, ''Fisherman should try the effects of the lobster trap on the Banks. The seal-trade, too, ought to be better attended to now, as gas-lighting has become general in the luxurious world.''[3]

MacTaggart's suggestion about seals and gas-lighting was astute. By the 1820s, Europe was experiencing a serious lighting crisis. Wars and limited supplies continued to push up the price of whale oil and Russian tallow at a time of ever-increasing demand. Fire-prone factories – particularly cotton mills – were searching for a safer and more economical source of light than candles or liquid-fuelled lamps.

The abundant body fats and oils of the seal provided a potential solution. These could be heated in a cast iron retort to produce gases for lighting and cooking. The generation of artificial illuminating gases was one of the most exciting and lucrative areas of advanced technology. MacTaggart, then, was suggesting that Canadian resources be combined with the most recent and promising technology to solve pressing problems and to improve the quality and safety of life.

John MacTaggart's work on the Rideau Canal provides, however, the best examples of his original approach to engineering in Canada. As a civil engineer on a wilderness project, MacTaggart experienced the frustrations of trying to establish elevations, water depths, and drainage patterns in impenetrable cedar swamps and dense forests. It was virtually impossible to take levels by conventional surveying techniques that had been developed in a mature agricultural landscape and depended on uninterrupted lines of sight.

His solution was ingenious. He put a "candle . . . in a small lantern on the index of the levelling staff" and then looked through his theodolite backwards. He found that "whenever the little star, formed by the candle on the staff is caught by the leveller looking through the object-glass, and the 'halt halloo' given, the level of the intervening land may be obtained with considerable niceness, and save an immense deal of labour in clearing brushwood and trees. By this method levels may be run through a wide tract of wilderness in a short time;

and when these levels come to be proved . . . they are always found to be wonderfully near the truth."[4]

MacTaggart took surveying equipment from one milieu, added other elements – the candle and the lantern – and then employed them in a distinctly original way. Using his method, the surveyor looked through the instrument backwards to obtain a larger field of view with a lower but acceptable level of accuracy. A simple trick, no new equipment, yet elegant and, above all, effective. And having solved the immediate problem, MacTaggart went on to suggest that surveyors sent aloft in tethered balloons would be an answer to Canada's long-term surveying needs.

MacTaggart's imaginative use of available resources to solve specific problems was ingenious and discerning. But his most significant contribution to North American engineering was his entirely new approach to the construction of the Rideau Canal. It is difficult to separate the relative contributions of By and MacTaggart, but it appears that the latter understood most clearly that new circumstances demanded new solutions. "The plan," MacTaggart wrote, "so far as I am aware, is new and has never been tried before; but the situation of the place, and many other circumstances, justify the method proposed."[5]

The Rideau Canal represents a fundamental re-evaluation of the idea of what constitutes a canal and how it should be built. Because of the relative scarcity of land, European canals consisted primarily of locks and excavated channels. The alternative to digging – whether for new or existing channels – as a means to make the water deeper is to erect dams to flood an entire area such as a valley. This method floods a larger area than is actually needed as a navigation channel.

The Erie Canal, the first monumental canal project in North America, followed the European approach of excavating and was widely praised by contemporaries

Far from major industrial centres and with often limited economic resources, Canadians in the nineteenth century developed a tradition of improvisation and innovation. A good example is shown in this photograph taken about 1870 at Grant's Sawmill, Matapedia Lake, Quebec. Here, a stationary locomotive engine was used as a power source for the mill, running a belt drive for a circular saw. The long stack was a further modification to prevent sparks starting fires. (Public Archives Canada C-6354)

as a marvellous engineering feat. But MacTaggart viewed it as deficient. Instead, he championed a new approach to canal building which would reduce construction and maintenance costs and take advantage of the relative abundance of land.

Describing the new approach, MacTaggart wrote: "It is not ditched out or cut out by the hand of man. Natural rivers and lakes are made use of for this Canal, and all that science or art has to do in the matter is, in the lockage of the rapids or waterfalls, which exist either between extensive sheets of still river water, or expansive lakes. To surmount this difficulty dams are proposed . . . by which means the rapids and waterfalls are converted into still-water." He went on to point out that "the extensive utility of these dams must be obvious to any person who considers the business in an engineering point of view; they do away with lines of extended excavations through a thick-wooded wilderness. In several instances, a dam not more than 24 feet [7.3 metres] high, and 180 feet [54.9 metres] wide, will throw rapids and rivers into a still sheet above it for a distance of more than 20 miles [32 kilometres]."[6]

MacTaggart also explained that his vision of canal construction dispensed with embankments which often required extensive repairs, and increased the total surface area of water in the system, thereby reducing the severity of flooding from a given amount of precipitation or melt water.

With John MacTaggart and the Rideau Canal, one can see how pre-1887 engineers were learning to be Canadian, to deal with the particular circumstances created by a vast wilderness country with a harsh climate. One can also see how the world was beginning to benefit from the special engineering knowledge and strengths born out of the Canadian experience.

Engineers Help Build a New Colony

The structures created by engineering embody the goals, values, and ideals of a society. Engineers fashion much of the built environment that defines the quality and nature of life, that expresses the culture and shapes the outlook of a people. In an emerging society, engineers are often the vanguard, defining a framework that will mould future development. For a few critical years in the early history of British Columbia, engineers were used intentionally by the British to extend and defend their civilization.

The discovery of gold in the mid-1850s in what was then called New Caledonia focussed world attention on the Pacific Coast of Canada. Thousands of outsiders scrambled to a neglected British possession administered by the Hudson's Bay Company as a source of furs, minerals, and timber. The largely American influx of miners, prospectors, and assorted get-rich-quick schemers threatened the recognition of the Pacific West Coast as a British possession.

To counteract this threat, Britain established the colony of British Columbia in the fall of 1858 under the governorship of James Douglas, known today as the father of British Columbia. As a longtime employee of the Hudson's Bay Company, Douglas knew a great deal about the land and the people he was to govern. To preserve British influence, Douglas pointed out that, as well as civil government and law enforcement, the new colony needed the underpinnings of settlement: mechanisms to raise and collect revenue; construction of roads, trails, and bridges; and exploration and development, including the charting of settlement patterns and townsites, and the surveying and mapping of land.

Sir Edward Bulwer Lytton, the British Secretary of State for the colonies, responded immediately to Douglas's request. By the end of 1858, Lytton had sent 150 non-commissioned officers and men in a company

The Alexandria Bridge, a notable early Canadian suspension bridge, was constructed over the Fraser River near Spuzzum, British Columbia, in 1863 as part of the Cariboo Road. This ink drawing by E.T. Coleman was done at the time of construction and shows the timber towers and wooden trusses which supported the bridge's 90-metre span. The bridge was, however, destroyed in 1912 to prevent Grand Trunk Railway employees from visiting the saloons and brothels in nearby Spuzzum. (Provincial Archives of British Columbia pdp 168)

of Sappers and Miners to British Columbia under the command of an officer of the Royal Engineers, Colonel Richard Clement Moody.

As a Royal Engineer, Moody was a member of an elite group, trained to meet both civilian and military needs. Most officers of the Royal Engineers were graduates of the Royal Military Academy and, unlike officers in other areas who bought their commissions, the initials "R.E." were earned through intelligence, accomplishment, and self-discipline.

From among the many volunteers for the company of Sappers and Miners, Lytton selected the 150 he judged best able to form the basis of a cultivated community in a newly formed British colony. Because they would be in an emerging society far from England and its influence, those chosen possessed a wide range of useful trades other than soldiering.

The impact of the Sappers and Miners was felt immediately. In the summer of 1859, the company surveyed and laid out the town of New Westminster, the temporary capital of British Columbia. The military engineers also surveyed the towns of Yale, Hope, and Douglas.

The most pressing need was, however, a road to the interior. The Cariboo Road which cut through rugged and inhospitable mountains is one of Canada's greatest early engineering achievements. During the project, civilians were hired to make up for a shortage of military engineers. The Sappers and Miners were given supervisory roles and assumed total responsibility for the most difficult section, some 10 kilometres along the banks of the Fraser River. Here the entire road had to be hewn from the side of nearly vertical cliffs, cut through rock, or built on piles. This extremely difficult piece of work added both to the experience and to the reputation of Canadian engineers, who were proving themselves to be masters of construction under challenging conditions.

The opportunity to work alongside skilled military engineers on the Cariboo Road provided a rigorous training ground for civilians in a land that lacked formal educational facilities in engineering. Joseph W. Trutch, an Englishman who had worked as an engineer in the United States, was a contractor on the Cariboo Road. With the completion of the Alexandria Bridge, a 90-metre suspension bridge spanning the Fraser River at Spuzzum, Trutch earned a place as one of the great pioneers of Canadian bridge building. In recognition of his work, Trutch was made Commissioner of Lands and Works, and then first Lieutenant-Governor of the new province of British Columbia when it joined Confederation in 1871. A number of other prominent and influential engineers – including Walter Moberley of CPR fame – also developed their engineering skills on the Cariboo Road.

The Sappers and Miners profoundly influenced the emerging society of British Columbia. They erected many of the buildings that were to form the social capital of the new communities, including schools, churches, a library, a hospital, and an observatory. They designed the first coat of arms, and even a postage stamp. Moreover, their engineering was in many ways unusually far-sighted. Colonel Moody insisted on integrating allowances for piped water and sewers into the original layout for a city, rather than letting the community grow until fire, disease, and strangulated growth forced the recognition of these needs. Moody also made the enlightened decision to preserve for Vancouver a tract of the original forest, to be known as Stanley Park.

In 1863, when Britain asked its colony to assume the costs of maintaining the military engineers, Governor Douglas refused and the company was disbanded. Rather than moving to another posting, 130 of the 150 engineers stayed on as civilian settlers, providing strong foundations and leadership in the settlement and industrialization of British Columbia. In the words of Frances Woodward, the historian primarily responsible for our understanding of the Sappers' and Miners' contribution, "They went quietly about their business of 'civilizing' the rough terrain."[7] To this day, British Columbia and particularly its capital, Victoria, retain a strong British influence.

The Welland Canal: Hard-Won Lessons

Niagara Falls, one of the world's greatest cataracts, located on the Niagara River between Lake Erie and Lake Ontario, presents a formidable obstacle to navigation. Each of the four Welland Canals – three built during the nineteenth century – were designed to circumvent this barrier and to link the Upper Great Lakes and inland areas of both Canada and the United States with the Ontario–St. Lawrence region and thereby to provide access to the Atlantic and thence to the rest of the world. The need for such a connection became critical in 1825 with the opening of the Erie Canal. Through the port of Buffalo on the Niagara River, the Erie Canal gave Lake Erie direct shipping access to the Atlantic entirely through United States territory, thereby threatening economic development in Canada.

Narrowly conceived as an answer to this U.S. threat, the Welland Canal, with its problem-plagued history, prompted the development of a new approach to the funding and staffing of engineering projects and helped greatly to strengthen the Canadian engineering profession.

Whereas the Rideau Canal was a defence project generously funded by the British government and built by Royal Engineers, the Welland Canal began as a purely commercial venture relying on private capital and civilian engineering. At no time was funding secure or adequate. As a result, although one can find examples of brilliant engineering, the project was troubled from the beginning. The canal route was never properly surveyed and the implications of its construction were not fully understood. To keep engineering and construction costs down, plans were changed continuously.

Moreover, even by North American frontier standards, wood was used too frequently and often inappropriately. The Welland was a ship canal, not a barge canal, and needed to be built more solidly. It was plagued by costly maintenance problems which were so serious that the canal was literally falling apart in the middle before the ends were completed. Perhaps worst of all, the locks were simply too small to handle existing traffic, let alone the even bigger ships then being introduced.

Despite its shortcomings, however, the Welland Canal spawned engineering and industrial complexes that were far larger and more sophisticated than those on other Canadian canals. The First and Second Welland Canals were designed so that excess water passing through canal raceways could be used as a source of industrial water power (as was done on the Lachine Canal in Montreal). As a result, the Welland Canals brought into being flour, paper, and cotton mills, breweries, shipyards, drydocks, and many other industries.

With the coming into force of the Act of Union of Upper and Lower Canada in 1841, the new Province of Canada took over the floundering First Welland Canal. This takeover, and the subsequent building of a new canal, represents not only an important development in the history of Canadian commerce and transportation, but also a significant forward step in Canada's relations with its engineers.

In 1837, the Upper Canada government had hired "two scientific and practical Engineers," Hugh Nicol Baird and Hamilton Killaly, to analyze the existing canal and make recommendations. Because of the troubled history of the Welland Canal and the successful completion of the Rideau Canal, one might expect that Baird and Killaly would recommend that major engineering projects be handled by military or government engineers rather than civilians.

The consulting engineers, however, recommended a mixed system of non-military civil servant engineers and civilian engineers employed as needed. The consultants' conclusion apparently stemmed in large part from the experiences of the most influential of the two, Hamilton Killaly, in his native Ireland.

Killaly had been an engineer with the Board of Works in Ireland, but grew increasingly frustrated with its chairman's policy of hiring his fellow officers of the Royal Engineers instead of civilian engineers. Generally,

Railway building made steel rolling mills an integral part of the early industrial development of Canada. The Toronto Rolling Mills are shown in this 1864 pastel drawing by William Armstrong. Armstrong was a trained engineer who often depicted industrial subjects. In the right foreground is a steam hammer, one of the most impressive and awe-inspiring nineteenth century machines. In the centre background, a group of men operate the rolling mill. (Metropolitan Toronto Library T-10914)

Completion of the Intercolonial Railway linking Central Canada with the Maritimes "with all practicable speed" was a condition of the British North America Act of 1867. This photograph shows the construction of an earth trestle on the Intercolonial Railway near Higgin's Brook, Nova Scotia, in August 1871. Sandford Fleming, chief engineer on the railway, often used earth embankments instead of bridges. This was both an economy measure and sound engineering practice. The earth and rock from one cut could be used as fill in nearby locations. (Public Archives Canada C-17698)

This group photograph of a survey team on the Intercolonial Railway was taken in about 1870 on the Tartigou River in Quebec. Each man holds an instrument or tool appropriate to his role. W.F. Biggar, the chief surveyor, stands in the centre with his transit and notebook, with a rodman to his right. Another surveyor is at the far right, while his rodman is in the left foreground. Axemen, an essential part of any Canadian survey team, stand with axes in hand or at their sides. (Public Archives Canada C-17695)

Traditional skills and techniques were used in building this 6.1-metre tall arch culvert over the Black River, Nova Scotia, on the Intercolonial Railway in August 1871. For example, hand and animal power were exploited to operate simple mechanisms and move massive stones. Stone masons built the falsework for arches and dressed the stone with such details as the fine edging of the voussoirs and keystone. (Public Archives Canada PA-21994)

The precise finish and beauty of materials of much nineteenth century construction have led to the assumption that every component, visible or not, was finely crafted. The carefully worked exterior stones on this culvert on the Intercolonial Railway might give such an impression. However, the photograph, taken about 1870 during the culvert's construction, reveals that economy required a strict distinction between those stones that were purely structural and those both structural and decorative. (Public Archives Canada PA-22137)

Royal Engineers worked on a project until its completion and then were assigned elsewhere. Killaly felt that the Board's practice of hiring Royal Engineers worked against the best interests of Ireland as it did not promote the development of an indigenous pool of engineering talent.

Dissatisfied, Killaly emigrated to Upper Canada. After co-authoring the report on the Welland Canal, he became engineer for the Welland Canal Company and, in 1841, chairman of the Board of Works of the Province of Canada. Killaly went on to occupy a number of other key public service positions, and he designed the Second Welland Canal which opened in the early 1850s. In whatever public position he served, however, Hamilton Killaly continued to play an important role in directing projects to civilian engineers who could add to the collective strength of Canada's engineering base.

The Welland Canal's problem-ridden history and Killaly's report and subsequent influential role in the project led to the emergence of a particularly Canadian tradition of engineering. It was recognized that some projects were either too big or unsuitable for private enterprise. But it was also decided that a system using mixed engineering talent, rather than purely governmental or military sources, would best serve Canadian needs. Under a mixed system, government funding and engineering expertise would be available for projects when they were required. At the same time, a mixed system gave private engineers employment and fostered an expanding indigenous tradition, industry, and profession.

The lessons of the First and Second Welland Canals were put into practice during construction of the Third Welland Canal which opened in the early 1880s. Civilian engineers employed as civil servants designed the canal; construction and a great deal of the engineering work was carried out by the private sector. Reliance on a combination of government and private-sector engineering talent is a system which has served Canada well to this day.

Railroads and the Development of a New Professionalism

The crucial role that railroads played in Canadian history and the excellence of the engineering on these projects have been well documented. The driving of the "Last Spike" in 1885, which marked the completion of the transcontinental railway and fulfilment of the bargain that had brought British Columbia into Confederation in 1871, is one of the best-known events in Canadian history.

However, the central role railroads played in fostering professionalism in Canadian engineering is less well understood. Canadian railroads were chronically short of funds and sound technical decisions often gave way to economic and political considerations. In forcing engineers to confront questions of professional ethics and independence, the railroads stimulated the growth of a mature Canadian engineering profession.

Perhaps the best introduction to these complex questions is through the histories of two engineers, Sandford Fleming and Thomas Coltrin Keefer. Both Fleming and Keefer were connected with Canadian railroads and the development of an independent engineering profession.

Sandford Fleming was chief engineer on the Intercolonial Railroad linking Central Canada with the Maritimes. Under Section 145 of the British North America Act, two of the original partners to Confederation – Nova Scotia and New Brunswick – obtained an agreement that "it shall be the Duty of the Government and Parliament of Canada to provide for the Commencement within Six Months after the Union, of a

This wrought-iron superstructure bridge on the Intercolonial Railway, over the Miramichi River, New Brunswick, was manufactured by Clark, Reeves and Company of Phoenixville, Pennsylvania, in 1872. It is one of three prefabricated "pin connection" trusses that Sandford Fleming purchased from the American manufacturer. Such bridges were ordered by catalogue and built to customer specifications. A log boom has been slung between the piers to prevent timber from floating further downstream where the river widens considerably. (W. Williams/Public Archives Canada C-77821)

Chief engineer Sandford Fleming protested strongly against the use of wooden bridges on the Intercolonial Railway. Eventually, he won his case after petitioning Prime Minister John A. MacDonald and then the Privy Council in Britain for more expensive, but more permanent, iron superstructures. The bridges consisted of plate and lattice girders set on stone piers, and were lean and elegant. This one at Trois Pistoles, Quebec, was photographed about 1870. Railways could not follow the rolling contour of the land – hence the high level of this bridge compared with the timber bridge in the background intended for horse-drawn vehicles. The two men crossing the bridge on a hand cart are dwarfed by its immense scale. (Public Archives Canada PA-22068)

The manufacture and maintenance of railway rolling stock and equipment fostered the growth of a wide variety of support industries in nineteenth century Canada. This 1884 photograph of the Canadian Pacific Railway machine shop in Winnipeg, Manitoba, was taken prior to electrification and the provision of a separate motor for each machine. Here, power is supplied from a central source and distributed by a drive train consisting of overhead line shafts, pulleys, and belts. (Notman Photographic Archives 1586)

At a length of 2742 metres, the Victoria Bridge over the St. Lawrence River at Montreal was one of the outstanding landmarks of nineteenth century engineering in Canada. The Grand Trunk Railway began work on the bridge in 1854. This photograph of October 25, 1858, was taken by William Notman, an eminent Montreal photographer hired to record the construction of the bridge. The piers of the Victoria Bridge had to be wide and tall enough to accommodate river traffic. They flared out on the upstream side at water level so that ice would ride up, break of its own weight, and float away. This reduced the risk of dangerous ice build-ups which could topple the bridge or cause flooding. (Notman Photographic Archives 7526)

At least three Trevithick locomotive engines are under construction in this 1860 photograph of a typical mid-century factory at Point St. Charles, Quebec. Although the rail industry was heavily dependent on iron and Bessemer steel, wood remained a plentiful building material. Here, for example, both the heavy ceiling beams and the locomotive cab are made of wood. (National Library 14849)

Railway connecting the River St. Lawrence with the City of Halifax in Nova Scotia and for the Construction thereof without Intermission, and the Completion thereof with all practicable Speed."

Fleming's work on the Intercolonial Railroad provoked one of Canada's most highly publicized and historically significant client-engineer disputes. Fleming insisted that clients – even those less than fully informed – be given the best value for funds expended. He championed the use of iron bridges on the Intercolonial Railroad right from the beginning, rather than as phased-in replacements for timber bridges and trestles.

In opting for more permanent structures, Fleming was flying in the face of conventional wisdom. His well-placed opponents claimed that timber was as safe as iron and considerably cheaper. Fleming countered that Canadian timber bridges burned with alarming regularity and that, given the route's proximity to navigable waterways, iron was as cheap or cheaper.

Overruled by his colleagues, Fleming first appealed to the Prime Minister of Canada and then to the British Privy Council where he finally won his case. However, his triumph in the cause of professionalism in engineer-

ing was in many ways atypical. Fleming possessed the advantages of position and fame, and he was able to employ these advantages in pursuing his case. Other, less well-placed engineers were not as fortunate.

The situation confronting average engineers in the late nineteenth century was accurately described by Thomas Coltrin Keefer, a Canadian-born engineer, in a series of lectures on civil engineering at McGill University in 1855–56. At the time of the McGill lectures, Keefer had a number of major engineering projects behind him, had published several highly acclaimed books, and was recognized as a leading engineering writer. He was a firm believer in the promise of modern engineering and particularly of railroads. In *The Philosophy of Railroads*, a highly acclaimed work published both in French and in English, he had written that "steam has exerted an influence over matter which can only be compared with that which the discovery of printing has exercised upon the mind."[8]

Yet in his McGill lectures, Keefer spoke eloquently and directly about the crisis in Canadian engineering and about the shame and "disgrace" which he felt was engulfing the profession. He told his audience, "The engineer, though an indispensable agent, is generally a junior partner in the firm of Grab, Chisel and Co."[9]

Keefer felt that economic pressures often forced engineers to act on behalf of questionable financial and political interests. He maintained that if engineers are to serve society as independent professionals, they must achieve a level of self-sustaining independence and respect similar to that enjoyed by lawyers and physicians. He advocated the professionalization of engineering, particularly through rigorous standards of education and training.

After his McGill lectures, Keefer went on to achieve further prominence both through his writing and through engineering projects. He became famous for his pumping stations and water supply systems in various major Canadian cities including Hamilton, Ottawa, and Montreal.

Keefer's major legacy to Canadian engineering was, however, his advocacy and activities on behalf of professionalism in engineering. He believed that if Canadian engineers were to serve the public good rather than narrow private or political interests, they required increased status, more self-control, and a clearer public identity. As a founder and first president of the Canadian Society of Civil Engineers in 1887, he put these ideas into practice. Keefer was also the first Canadian president of the American Society of Civil Engineers, president of the Royal Society of Canada, and an honorary member of the Institute of Civil Engineers of Great Britain.

As the historian H.V. Nelles observed: "Keefer personified the best qualities of the engineering profession. An urbane, erudite, physically attractive figure, he was one of those individuals whose wide personal esteem brought welcome dignity to an emerging profession."[10]

With the growing influence of such noteworthy figures as Fleming and Keefer and the founding of the Canadian Society of Civil Engineers in 1887, Canadian engineering was emerging from an era of dependence and reliance on European traditions to one of increasing maturity and professionalism. This professionalism was to be tested and proven repeatedly in the coming years, as the young nation moved forward confidently on a wave of industrial expansion into the twentieth century.

2

Building on a Legacy of Achievement: 1887 to 1900

The completion of the transcontinental railway in 1885 and the formation of the first national engineering society, the Canadian Society of Civil Engineers, in 1887, marked Canada's transition into the ranks of advanced technological and engineering nations.

According to its original by-laws the Canadian Society of Civil Engineers was to facilitate the acquisition and the interchange of professional knowledge among its members, and to encourage original investigation. Equally clear, but unstated was the goal of ensuring that engineers received proper recognition for the contributions they were making to the country. Late nineteenth-century Canadian engineers realized that the best basis for reaching these goals would be a solid record of engineering achievement. Building on the legacy of earlier generations of Canadian engineers, they combined remarkable accomplishments in such traditional areas as railroads and canals with innovative applications of the emerging science of electricity.

Engineers Establish a Professional Association

The situation of engineers in late nineteenth-century Canada was highly contradictory. On one hand, they were acknowledged as the vanguard in transforming a vast uncharted land into a modern urban-industrial nation. On the other hand, engineers possessed little job security, and had far lower status and remuneration than those in comparable professions such as physicians and lawyers.

Victorian attitudes tended to reinforce the importance of engineers. Victorians saw the earth and its riches primarily as something to be explored, understood, subdued, and rendered productive; these ambitions could not be satisfied without engineers. Moreover, the engineers' own view of their role in realizing the enormous potential of a virgin land further aroused their enthusiasm and sense of mission.

Yet, despite the self-confidence that engineers might have felt as individuals, and despite their recognized value generally, the profession as a whole was disorganized and its members felt themselves vulnerable and unappreciated.

A central problem was the almost complete lack of professional standards. Given the political origins of many projects, work tended to be judged on political rather than technical criteria. Engineers were sacked for speaking out about shoddy design and workmanship, or for merely entertaining the wrong political loyalties. With no standards for entry into the profession or for disciplining unprofessional conduct, virtually anyone could be designated an engineer and be given full professional responsibility and respect.

An underlying cause of the profession's lack of standards, disorganization, and low status was its relative infancy. Civil engineers – that is, non-military engineers – were in effect a nineteenth-century invention. By contrast, law and medicine, with centuries of history and a respected educational system, were self-governing and self-regulating monopolies, with the legal power to decide who could and could not practice. Furthermore, whereas medicine and law were regarded as acceptable professions for the upper classes, engineering was viewed as somewhat of a trade and was filled largely from the ranks of farmers, small merchants, and tradesmen.

The Canadian Society of Civil Engineers was formed to answer the recognized need among engineers for increased status, more self-control, and a clearer public identity. What is perhaps most surprising about the society's formation is that a number of influential engineers opposed the establishment of a legal monopoly with licensing and disciplinary powers similar to those found in medicine and law associations.

Such prominent spokesmen as Thomas Coltrin

These two photographs present nineteenth century engineers in two of their major roles. In the first, taken about 1870, a survey party led by W.F. Biggar, near the Dry Tartigou River in Quebec, is dressed for the field. In the second, formally dressed engineers stand proudly in front of the Montreal headquarters of the Canadian Society of Civil Engineers at its annual meeting in 1894. (Public Archives Canada C-25624, PA-148562)

Keefer, the society's first president, and John Kennedy, its vice-president, viewed the acquisition of professional knowledge as the key to enhanced status and recognition. They therefore wanted a professional organization whose aim would be primarily educational.

Keefer and Kennedy were opposed by a group led by Alan MacDougall, a Toronto consulting engineer and the society's provisional secretary. MacDougall agitated to have engineering become a closed, legally defined profession.

At the society's first general meeting on February 24, 1887, Vice-President Kennedy made it clear that MacDougall's faction had lost. Plans to acquire legal rights had been abandoned, Kennedy reported, and the profession of engineering would remain open and "accessible to those in whom the public had confidence."[11] The society would provide and encourage educational opportunities to elevate the knowledge and utility of those in the profession. In addition, it would act as a force for research in engineering.

The decision to abandon the idea of a professional association in favour of what today would be called a learned society reflected popular Darwinian ideas about the survival of the fittest as applied to human society and institutions. The best engineers – those who qualified for membership in the society – would thrive, while the incompetent and dishonest would find themselves without work. The decision also reflected the Victorian faith in progress and in education as a means of improving both the individual and society.

To ensure its reputation as the new, technologically informed elite, the Canadian Society of Civil Engineers imposed stringent entrance requirements. Engineers had to be at least 30 years of age with a minimum of 10 years of combined work and educational experience. They required at least five years of what was called "responsible charge of work." In addition, they were to have the respect and confidence of their fellow professionals as demonstrated through either a vote in a letter ballot or letters of recommendation. Equally important, each engineer was to be a cultivated, well-rounded gentleman such as the first president Thomas Coltrin Keefer.

Although the decision to structure the new organization as a learned society was perhaps the most workable course under the circumstances, it had significant drawbacks. Because the profession remained open, employers could pick whomever they wanted and anyone could be styled an engineer. Members were not protected from competition with the untrained and incompetent. Nor were there any effective legal, financial, or moral sanctions to guard engineers from being treated unfairly by unscrupulous employers.

There were also complaints that the society was controlled by a small group of older and more conservative engineers, known as the McGill or Montreal clique. In fact, the first four secretaries of the society between 1887 and 1925 all held teaching positions in the Faculty of Applied Science at McGill University. The dominance of this group did, however, reflect Montreal's importance as an engineering centre. Furthermore, were it not for the active work of prominent Montrealers, the society would certainly not have gained as much public prestige as it did, and it might not have survived.

Despite the crucial importance of Montreal, the society was in fact as heterogeneous as Canadian society and the engineering profession. From its inception, officers were drawn from French, English, and other linguistic and cultural groups. The first two presidents were Ontario born, the next Russian born of Polish descent, and, in 1905, Ernest Marceau, head of the Ecole Polytechnique in Montreal, became the first Quebec-born, francophone president.

Charles F. Baillargé was another prominent member. Perhaps one of the most respected and accomplished of nineteenth century engineers, Baillargé embodied the excellent education and strict adherence to principles – even at the cost of his own employment – that characterized the best of his fellow engineers. Baillargé dedicated his life to promoting public awareness and engineering education.

The Canadian Society of Civil Engineers promulgated an ideal of professionalism in engineering that was perhaps best expressed in the presidential address of P. Alex Peterson. Peterson wanted his fellow engineers, particularly the younger ones just entering the profession, to understand the importance of economic factors in Canadian engineering. He explained that good engineering – that is, economical engineering – attracted investment capital and thereby played an important role in national development, prosperity, and the continuing employment of engineers.

Peterson's example was the Canadian Pacific Railway. European railroads were built as finished structures and were intended to last for decades. In Canada, significant economic advantages were to be gained by using temporary construction. Temporary work reduced costs and generated revenue much faster by allowing traffic to be carried earlier. Later, temporary structures could often be incorporated into more permanent ones.

In concluding his address, Peterson articulated an engineering philosophy that stressed effectiveness and integrity above all else.

It is quite an easy matter to build an expensive structure, but it is an engineer's duty to build an effective structure for the least possible cost, and after his design is made perfect as to stability, he should proceed to remove from it everything that is not absolutely necessary and that has no duty to perform, remembering that he must never build ornaments, but that good and wise construction will be ornamental in itself.

Finally you must also remember that your success in life depends on your capacity and willingness to take infinite pains with everything you are called upon to carry out. You must be in downright earnest about your work, and, above all things, you must be absolutely and entirely honest in every aspect, never letting your convictions or opinions be warped in any way for any consideration, and then, if you may not always command success you will at least deserve it, which is often better. [12]

Peterson and his colleagues at the Canadian Society of Civil Engineers clearly recognized that, if engineering was to emerge as an independent profession capable of serving the needs of society rather than narrow political or economic interests, it needed both improved standards and societal recognition of its utility and importance. Organization was one key to improved standards and status. Another was achievement.

Innovations in Tunnel Building

By 1887, Canadian engineers had built a transportation system of railroads and canals unequalled in the world. But this singular accomplishment had brought with it many problems, one of which, the crossing of navigable waters by the railroad, prompted the construction of the remarkable St. Clair Tunnel.

Linking Sarnia, Ontario, with Port Huron, Michigan, the St. Clair Tunnel was an unprecedented achievement in tunnel building and helped establish the reputation of Canadian engineers as world leaders in the field.

The tunnel was built as part of the Grand Trunk Railroad, a Canadian and British-owned line which ran from Chicago through Canada and on to Portland, Maine. The Grand Trunk was one of the continent's busiest railroads. A steady stream of agricultural products – particularly wheat and meat – augmented by ever-increasing numbers of manufactured goods and passengers flowed steadily in both directions.

However, at the Port Huron–Sarnia crossing, the long trains had to be broken up to cross the St. Clair River by car ferry. Time was lost in disassembling and reassembling trains and in loading and unloading the ferry. There were also problems of river ice. These inconveniences and dangers, combined with a steady increase in traffic, produced backlogs on both sides of the river.

Sir Henry Tyler, president of the Grand Trunk Railroad and a former Royal Engineer, commissioned Walter Shanly, one of Canada's foremost engineers, to do a feasibility study of both bridge and tunnel crossings.

Shanly had three alternatives to consider: a fixed bridge, a movable bridge, or a tunnel. A fixed structure

Wooden railway trestles, such as this one over Mountain Creek, British Columbia, circa 1880, were regarded as one of the wonders of late nineteenth century Canadian engineering. Wood was used extensively because it was both plentiful and relatively inexpensive. However, wooden structures such as this bridge would later be replaced with structures made from more permanent materials. (Public Archives Canada PA-66576)

A project such as the St. Clair Tunnel between Sarnia, Ontario, and Port Huron, Michigan, was a testament to the economic importance of exchange between the two countries. These sketches describe the construction of the tunnel. They first appeared in *Scientific American* in 1891 and were later published by *The Dominion Illustrated*. The map shows three major railway lines, two from the U.S. and one from Canada, all converging on the bottleneck area of Port Huron and Sarnia. The sketch below shows a section of the St. Clair River where the tunnel was cut through a layer of soft blue clay. Brick bulkheads with air locks were built across the tunnel openings at each side of the river to keep the air pressure inside the tunnel at acceptable levels. (Public Archives Canada C-129822)

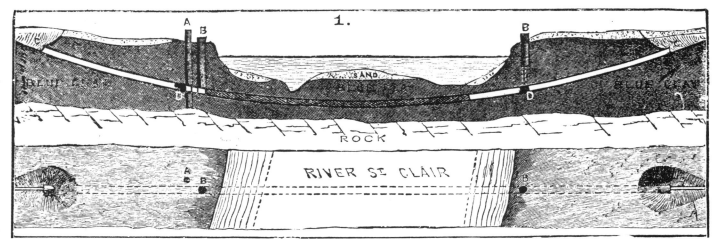

required long approaches to get land vehicles to a height that would permit the passage of ships. A movable bridge, on the other hand, would slow down both land and water traffic because, at any given time, it restricted movement to either one or the other. Although a tunnel avoided these problems, it was dangerous to build and, from an engineering viewpoint, far more challenging.

Nevertheless, Shanly recommended a tunnel and, in 1884, the Grand Trunk Railroad created Canadian and U.S. subsidiaries, the St. Clair Frontier Tunnel Co. and the Port Huron Railroad Tunnel Co., respectively, to work from each side of the border.

The task of transforming Shanly's concept into a workable tunnel fell to project engineer Joseph Hobson. Born in 1834 in Guelph, Ontario, Hobson apprenticed as an engineer in Toronto. After working in railway construction in the United States, Ontario, and Nova Scotia, he became resident engineer of the Grand Trunk Railroad in 1873. Two years later, he was appointed chief engineer.

To keep the railway grade to a minimum while at the same time avoiding the cost of long approaches, Hobson decided to minimize the depth at which the tunnel would lie beneath the riverbed. A thick layer of blue clay overlay bedrock under the river, and if one could tunnel through the clay, there would be no need for the slow, dangerous, and costly process of blasting through bedrock.

It was a daring plan. Tunnelling was started from both sides of the St. Clair, but the aggregate lengths had scarcely exceeded 61 metres when quicksand, gas, and water began seeping into the bore faster than they could be pumped out. Hobson was forced to halt construction, backers lost confidence, and the project faced a financial crisis.

Although new financing was arranged, no independent contractor would risk taking on the job and Hobson was forced to go it alone. The ultimate success of the project was due to Hobson's original and technologically advanced approach to tunnel building – his use of novel tunnelling shields, standardized prefabricated cast-iron casings, and compressed air.

Tunnelling Shields

Cylindrical shields to provide structural support and a platform for workers during construction had been used before. But Hobson designed two identical shields

These wood engravings of the construction of the St. Clair Tunnel occupied a full page in *The Dominion Illustrated*, October 10, 1891. Such prominence demonstrates that the tunnel was widely recognized as a significant engineering accomplishment. Here, installation of a prefabricated cast-iron lining follows hand excavation in three tiers in the tunnelling shield. (Public Archives Canada C-129823)

Hydraulic Tunnelling Shield

1. Rear view of the shield, showing hood and rams.　2. The shield in place on grade.　3. Interior view of shield and tunnel.　4. Front view of shield.
5. Lowering of the shield to the heading.

CONSTRUCTION OF THE ST. CLAIR TUNNEL.

of unprecedented size. Made largely of steel plate 2.5 centimetres thick, each shield was a 72.6-tonne cylinder, 4.9 metres long, and an extraordinary 6.6 metres in diameter – almost twice the size of earlier shields.

The shields started on opposite sides of the river, each one driven forward and kept in alignment by 24 independently operated rams. The shields gradually worked their way toward each other, their sharp edges cutting into the clay, which was removed by workmen with pick and shovel and loaded into tramcars pulled along rails by mules.

Designed by Hobson and made by the Tool Manufacturing Company in Hamilton, Ontario, the shields performed flawlessly, pressing forward at a monthly rate of 138.7 metres. When, after a year of tunnelling, the two shields met on August 30, 1890, they were only a few centimetres out of alignment. Furthermore, unlike most tunnelling shields, these became part of the tunnel lining after their working parts were removed.

Prefabricated Cast-Iron Casings

As tunnels move forward in soft ground like the clay of the St. Clair River, they must be lined to prevent collapse. Brickwork or masonry was the lining normally used, but Hobson opted for prefabricated cast-iron casings, as had other progressive engineers. When bolted together, these formed an iron lining ring which, in addition to casing the tunnel, also served as a mounting point for the hydraulic rams advancing the shield.

Each cast-iron segment weighed about 454 kilograms. In order to bolt it to its mate, it had to be positioned with great accuracy. To do this with precision and safety, Hobson designed an erector arm as an integral part of the shield – the first apparatus of this kind ever devised.

Compressed Air

Hobson's first attempt to drive the tunnel had failed when hydrostatic pressure forced water through the clay and into the tunnel bore. This time, he sealed the tunnel by building brick bulkheads with air locks across the tunnel openings at each side of the river. Inside the tunnel, air pressure varying from 68.7 to 192 kilopascals above atmospheric pressure kept water inflow at an acceptable level.

The procedure posed several dangers. Too much air pressure could cause a sudden rupture up through the river bottom. The subsequent rapid loss of air pressure would cause the tunnel to flood. Workers in the compressed air of the tunnel also risked nitrogen embolism or the "bends" – then little understood. Despite precautions, three men and several horses and mules died.

Nevertheless, Hobson's approach to tunnel engineering worked extremely well. Stretching 966 metres beneath the St. Clair River, the tunnel was opened to freight service on October 27, 1891, and to passenger service on December 7.

Hobson's achievement was widely praised; recognition went far beyond the local press and specialized engineering works. *The Dominion Illustrated*, one of Canada's most popular magazines, published a number of drawings, some pirated from the *Scientific American* in New York, and a statement from the *Toronto Telegram* extolling Hobson's virtues.

Hobson, the engineer, is one of the earth's useful heroes . . . . He not only sketched the outlines of

This model of the tunnelling shield for the St. Clair Tunnel, based on engravings that appeared in *Scientific American* and *The Dominion Illustrated*, is on view at the Smithsonian Institution in Washington, D.C. (The Smithsonian Institution 49260-D)

a gigantic enterprise, but invented new means of working out his ideas. His daring achievement is one that any country might be proud of. The tunnel is a triumph of Canadian genius, and the success of Joseph Hobson is proof that Canada does not need to import talent even to design or execute the greatest engineering works.[13]

The tunnel cut off two hours of travel time and saved the Grand Trunk Railroad $50 000 a year in ferry costs. It garnered a number of firsts: the largest and first successful major underwater tunnel in North America; the first in the world large enough to accommodate full-scale rail traffic; the first tunnelling shield built with an erector arm.

The most important feature of Hobson's work was, however, that it established and demonstrated the basic methodology for soft-ground tunnelling. Nearly a century later, the methodology remains essentially the same.

Nor was the St. Clair Tunnel an isolated case. For example, in terms of establishing and proving the fundamentals of tunnelling, the Hoosac Tunnel in Massachusetts was the hardrock equivalent of the St. Clair. Begun in 1851, the Hoosac was plagued by repeated failures until it was finally completed in 1875 by Canadian engineers Francis and Walter Shanly. (Walter Shanly had also been responsible for recommending construction of the St. Clair Tunnel.) In successfully completing the Hoosac, the Shanlys established new directions in hardrock tunnelling techniques that survive to this day.

The St. Clair Tunnel and similar projects such as the Hoosac stand as tributes to nineteenth-century Canadian engineering technology, manufacturing skill, and project management. Such achievements far surpassed anything one might have expected from a young, industrially developing nation with a limited population.

Electricity Transforms Canadian Industry

At the beginning of the nineteenth century, electricity was a mere curiosity studied by natural philosophers. Few foresaw its practical applications. By the end of the century, however, electric lights and electric motors were working a fundamental transformation in the way people lived and worked.

Most of the innovative work in electricity was done in Europe and the United States, but in adapting the new systems to local conditions, Canadian engineers made many fundamental contributions and rapidly assumed a position of leadership in the application of the new technology.

Modern electric lighting can be dated to 1876 when Pavel Jablochkov, a Russian engineer living in Paris, invented a much improved carbon arc light. A Gramme Dynamo-Electric Machine turned by a steam engine generated the current. Jablochkov's system provided a continuous direct current (DC) rather than the fluctuating currents of machines predating the Gramme ring. The light was more brilliant than anything else available and had neither the noxious odours of gas lamps nor their excessive heat.

The new electric lighting was introduced into Canada in 1878, when a French-speaking Montrealer, J.A.I. Craig, returned from the Paris Exposition where he had been

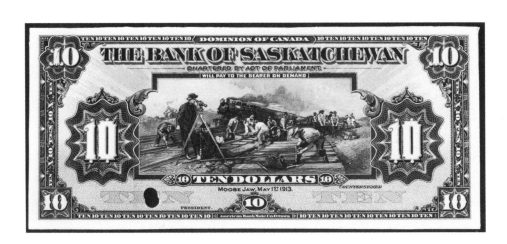

The vignette featured on this Bank of Saskatchewan $10 bill dated May 1, 1913, underlined the importance of technology and engineering in the development of the province. However, the Moose Jaw based bank never opened and the currency never went into circulation. (National Currency Collection, Bank of Canada)

Electrification and incandescent lighting changed the face of Canadian cities. This lamp post design, suitable for arc or incandescent lights, was included in a submission by Craig and Son to provide electric lights for the streets of Montreal. (City of Montreal Archives)

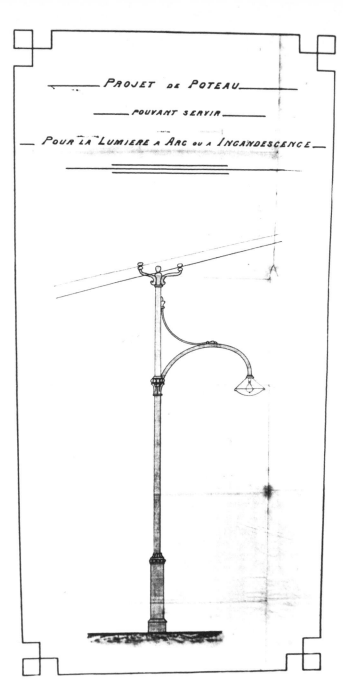

captivated by Jablochkov's electric candles. Craig set to work immediately and produced the startling demonstration recounted in *Québec: un siècle d'électricité.*

An autumn evening in Montreal, 1878. Only the faint glimmer of gas lamps lights the way . . . .

Suddenly, a bright flash illuminates the street. Amazement, then enthusiasm seizes the onlookers. Never before had such a brilliant lamp been seen, a light one hundred times more powerful than the gas jets along the avenue. The famous "electric candle" arc lamp, invented two years before in Paris by the Russian engineer Pavel Jablochkov, had been lit for the first time in Canada.

Electric lights were installed on high posts, 180 metres apart, at an early installation in Montreal Harbour in 1880 which was followed later by more sophisticated systems. This engraving appeared in *Picturesque Canada*, a national travelogue published in 1882, which reported: "All the modern appliances for loading and unloading are employed, and the facilities for almost immediate transshipment from freight cars to the hold of vessels are unsurpassed. . . . The result is continuous labour." (Public Archives Canada C-82834)

By 1895, the Royal Electric Company had
a large factory in Montreal manufacturing
parts for electric generators. This photograph of the shop floor illustrates the size
and complexity of the manufacturing industries that grew up to meet Canada's growing
electrical needs. (Hydro-Québec)

Montreal, then the nation, along with Europe and the United States – all were participating, thanks to electricity, in the most significant technological revolution since the discovery of steam-generated power. [Translation][14]

The transition from public curiosity to commercial commodity occurred quickly. Early in the 1880s, Craig and others formed a variety of small companies offering electrical services in Montreal. In 1884, the Montreal Harbour Commission purchased a Brush arc light system, named after its American inventor, Charles Francis Brush, and became the first public body in Montreal to electrify. In 1889, the Royal Electric Company signed a contract with the city to provide street lighting, thus marking a movement toward greater uniformity and concentration in the industry.

Other Canadian cities followed similar patterns of development. In Toronto, the arc light first appeared in a restaurant on Yonge Street where it was used, along with free ice cream, to attract customers. In 1881, the Toronto Electric Light Co. won the contract to light the streets of the city. The following year, visitors to the Toronto Exhibition marvelled at the four dynamos and 41 electric arc lamps set up for the occasion.

Progress in other centres was equally rapid. The large industrial complex on the Chaudiere Falls on the Ottawa River between Ottawa and Hull was using electric lighting for the mills as early as 1882. The following year, the Parliament Buildings were supplied with electric light from their own steam-powered generators. And, by 1888, Vancouver could claim to be the best lighted city of its size in the world.

Although the arc light was spectacular, newsworthy, and had many applications, it also had significant drawbacks. In arc lights, a spark jumped across a gap between the carbon rods. Illumination was provided by the burning, white-hot tips of these rods. However, the illumination was extremely bright – so bright in fact that arc lights were unsuitable for home use or any other application which did not involve illuminating a large, open space. In cities, the normal practice was to turn off the arc lights at midnight and use only gas lamps until dawn.

Arc lights were temperamental too, and needed constant maintenance. As the rods burned and increased the gap, one had to provide a mechanism to keep the gap constant or increase the voltage. The system was expensive as well. In Paris, for example, each set of 16 lamps required a Gramme Dynamo-Electric Machine powered by a 16-horsepower steam engine.

The arc light's intense brightness, costliness, and maintenance requirements made it inferior to gas lamps for mass public lighting and in many other applications. During the latter half of the 1880s, the incandescent light, with its softer, less intense illumination and virtual freedom from maintenance, began to oust both gas and arc lighting from most sectors of the marketplace.

The key contribution to incandescent lighting was made by the famous inventor, Thomas Alva Edison. Edison was the first to develop a complete system around the incandescent light bulb that could compete with gas for the low-cost, mass lighting market.

As part of the system, Edison's well-funded research laboratory produced a high-resistance bulb that could be fitted into a simple parallel circuit so that it could be turned on and off without interfering with other circuit elements. Other aspects of the design included a transmission system for the low-voltage alternating current (AC) required. Power could be drawn from either the customer's own generating system or a central supply station.

The first central station used by Edison was the famous Pearl Street Station in New York City which was put into operation on September 4, 1882. By 1884, it was supplying power to 11 272 lamps in 500 homes.

By the 1890s, myriads of privately owned telegraph, telephone, and electrical lines on poles were creating a chaotic situation. This photograph with the caption "Laying The Electric Wire Pipes Underground In Toronto" appeared on the front page of *The Dominion Illustrated* on January 4, 1890. Laying wire underground cleaned up the urban environment and ensured more reliable service free from the hazards of inclement weather. (Public Archives Canada C-96697)

At the same time, over five times as many lamps were powered by customer-owned generators.

The enthusiasm for AC incandescent lamps caught on quickly in Canada. Before 1890, private installations included the New York Life Insurance Co. and the CPR station in Montreal, and the Parliament Buildings in Ottawa.

Nowhere, however, was the trend to incandescent lighting more pronounced than in industrial installations. With high levels of explosive fabric fibres in the air, nineteenth-century textile mills were particularly prone to disastrous fires. The enclosed incandescent lamp was far safer than previous systems that had included open flames, exposed sparks, and heating elements.

Application of incandescent lighting to textile mills in Canada began in February 1883, when an Edison system was installed to power 500 lamps in the Canada Cotton Company's mill at Cornwall, Ontario. By 1893, this had increased to 1250, making it the largest installation in the country. Other early mills using incandescent lighting included the Montreal Cotton Company at Valleyfield, Quebec; the Stormont Cotton Company at Cornwall, Ontario; the Gibson Cotton Company, Marysville, New Brunswick; and the Magog Print Co., Magog, Quebec.

In a paper presented to the Canadian Society of Civil Engineers in 1890, Canadian electrical engineer A.J. Lawson summed up the rapid penetration of electric lighting in Canada during the previous decade.[15] Lawson, who had built nearly 30 per cent of Canada's electric lighting capacity, noted, "Ten years ago there was not one electric light generating plant in operation in Canada." By 1890, Lawson estimated that "out of a total of five and a quarter million horse-power developed by steam engines and water wheels on this continent, half a million horse-power, or nearly ten per cent, is used in the production of electric current for the distribution of light and transmission of power, and in the manufacture of electrical machinery and appliances."

The United States had more plants, Lawson said, but on a per capita basis Canada was holding its own and installations spanned the country. "At the present time there are 13 530 arc lights and about 70 765 incandescent lamps in use throughout the Dominion. There is hardly a village in Ontario, having a population of over three thousand inhabitants, which has not an electric light station of some kind in operation, and few of the important towns of the other Provinces are without electric lighting."

In contrast with many other countries, Canadian streetcar systems had to be designed to run in extremely cold, snowy weather. The sweeper shown here on the Ottawa Electric Railway in 1891 made year-round operation possible. Without it, cars such as the elegantly outfitted Duchess of Cornwall and York – shown below in front of the Ottawa Post Office and the East Block of the Parliament Buildings in 1901 – would have been unable to run for long periods during Ottawa's snowy winters. (Public Archives Canada PA-8420, C-26390)

The transition to electric lighting, which took place relatively quickly across Canada, represented one of the most significant contributions of Canadian engineering to the building of a modern industrial nation. The pursuit of scientific knowledge had become increasingly respectable, and Canadian engineers were emerging as the acknowledged authorities in the practical applications of science in this country.

A New Power Source – the Electric Motor

The electric motor was the other great electrical innovation in the late nineteenth century, and Canadian engineers demonstrated a swift awareness of the new technology and an avid desire to put it to its best use.

The earliest widespread application of the electric motor was in street railways. Canada's first electric streetcar line began operating at the exhibition in Toronto as early as 1884, only five years after the world debut of an experimental system at the Berlin Industrial Exhibition. However, these early electric streetcars were unreliable and underpowered – the Toronto cars, for instance, initially worked well only on the downhill portions of the line.

The successful commercial operation of electric railways required the development of a powerful electric motor that would perform reliably under adverse weather and load conditions. To free the streetcar from carrying batteries, the motor would also have to operate on power transmitted from a remote generating station.

A street railway system meeting these requirements was first demonstrated at Richmond, Virginia, in 1887 by a former Edison employee, Thomas Sprague of the Sprague Electric Railway and Motor Company. Sprague's motor supplied the necessary torque, lubricated itself, and was enclosed for cleanliness.

The system also demonstrated the feasibility of using an off-car power source. This included a power distri-bution and take-up system for the trolley cars, and a car power-controller capable of handling the heavy amperages required.

By 1892, Sprague had added a multiple controller that enabled one driver to control the power delivered to several cars. In this way, a number of cars could be connected to form a train.

Sprague's breakthroughs made electric streetcars commercially viable and a number of Canadian companies rushed into the business. By mid-1890, electric street railways were working in Windsor, St. Catharines, Victoria, and Vancouver. In 1891, the first electric tramway for mining was put into the pits of a coal mine at Nanaimo, British Columbia. The generators that powered the tramway also provided electric light in the mines. In the same year, Canada's first interurban electric tramway began running between Westminster and Vancouver.

Few Canadians believed, however, that electric street railways could provide winter transportation in areas of heavy snowfall. This belief was reflected in the contract between the Ottawa Electric Railway Company and the City of Ottawa. The contract specified that the company could revert to horse-drawn vehicles for up to three months during the worst winter snows.

The Ottawa Electric Railway Company started its electric cars in the summer of 1891 and was able to keep them going through the winter by pioneering the design and use of electric sweepers to clear snow and ice from the tracks. Two years later, the company made winter travel more convenient and comfortable when it began running the world's first regular service on electrically heated streetcars. Across Canada, other cities followed Ottawa's lead in providing year-round, comfortable, electric streetcar service.

Although electric streetcars marked the first widespread application of the electric motor, they represented only a small portion of the motor's broad range of potential uses.

Electric streetcars employed the DC motor which, because of readily adjusted speed over a wide range, was well-suited to transportation. On the debit side, the commutator, an essential part of DC motors, required considerable maintenance. A major breakthrough, which opened up many more applications, occurred in 1887 when Nikola Tesla introduced an entirely new type of electric motor, the AC induction motor.

The AC induction motor overcame the major drawbacks of its DC counterpart. It was more rugged and it turned at nearly constant speeds regardless of the load – an important consideration in many industrial applications. Because it could be made in all sizes, speeds, and voltages, the AC motor lent itself to more varied uses. It could be made both sparkless and watertight and was consequently suitable for operation under adverse and hazardous conditions.

Most important, the AC induction motor operated off alternating current which was becoming the standard for lighting applications. The new motor made it possible to supply both lighting and rotary power needs from the same system.

During the 1890s, the use of electric motors, spurred by the development of the AC induction motor, spread rapidly in Canadian industry. Electric motors supplied large amounts of finely controlled power for numerous applications in factories, mines, machine shops, foundries, forges, flour mills, and grain elevators.

In an address given to the Canadian Society of Civil Engineers in 1894, Fred A. Bowman summed up the importance of electric motors in industry:

The advantages of electric motors for use in driving the machinery in small industries are efficiency, reduced cost of attendance, cleanliness, reduced fire risk, and economy of power . . . .

Those who had never handled them do not realize what a well built electric motor will stand in the way of overload and general abuse . . . electric motors have passed the experimental stage and taken their place as thoroughly reliable machines.[16]

By the late nineteenth century, through these new developments in electrical technology which Canadian engineers had adapted with alacrity to local needs, and through the outstanding success of projects such as the St. Clair Tunnel in which Canadians had built on their established strengths in railway and canal building, the engineering community was winning national and world attention. In so doing it was fulfilling the goal set for it by the newly formed Canadian Society of Civil Engineers – to form a strong, independent, and progressive profession. And it was contributing with distinction to the emergence of Canada as a modern industrial nation.

3

Pioneer Engineering in "Canada's Century"

In an address to the Canadian Club in 1904, Prime Minister Sir Wilfrid Laurier proudly declared, "As the nineteenth century was that of the United States, so, I think the twentieth century shall be filled by Canada."[17]

"Canada's Century," it was called, and the phrase reflected the widespread optimism and confidence Canadians experienced at the turn of the century. Historian Richard Clippingdale describes the prevailing atmosphere during Laurier's tenure as Prime Minister from 1896 to 1911:

The years of his Prime Ministership were a time of triumphant, unprecedented, boisterous, awe-inspiring growth in Canada.

With this expansion of enterprise, wealth and numbers, there developed among countless businessmen, politicians, professionals, journalists, and writers an extraordinary sense of optimism about the Canadian future.[18]

To meet the needs created by the explosive growth of the population and virtually every sector of the economy, engineers were called on to supply transportation and all manner of industrial, urban, and rural structures, systems, and equipment.

The raw numbers vividly demonstrate the challenges confronting Canadian engineers. For example, in 1897 Canada received only 21 716 immigrants. The year 1901 marked the beginning of a period of unprecedented increase, with the arrival of 55 747. Two years later, the number had climbed to 141 465, and by 1911, this had reached 331 288. Immigration peaked in 1913, just prior to the First World War, at 400 870.

Cities swelled and entirely new ones came into existence. Between 1891 and 1911, Montreal grew from a city of 219 616 to one of 490 504 and Toronto from 181 215 to 381 833. Western cities expanded even faster. Between 1901 and 1911, Vancouver's population leapt from 29 432 to 210 487, Winnipeg's from 42 340

to 136 035, Edmonton's from 4176 to 31 064, and Saskatoon's from 113 to 12 004.

Economic indicators followed a similar trend. The acres of golden wheat on immigration brochures advertising Canada were matched by the reality of Canadian wheat production, which soared from eight million bushels in 1896 to 56 million in 1901, and reached 231 million in 1911. Between 1900 and 1910, the value of Canadian manufacturing increased from $214.5 million to $564.5 million.

The unprecedented growth in population and economic activity created opportunities in traditional areas of Canadian engineering expertise such as the development of urban facilities and systems, and the construction of transportation networks and equipment. Advances in science and technology, such as the invention and wider usage of Portland cement, opened up even more opportunities and challenges.

A series of dramatic and high-profile projects during the first years of the twentieth century established the leadership of Canadian engineers abroad and created widespread public awareness in Canada of their important contributions. Shawinigan Falls in Quebec, Canada's first large-scale hydro-electric power complex, is a prime example.

Shawinigan Falls Project Delivers Power to Montreal

On January 15, 1898, the Quebec Legislative Assembly issued a far-reaching corporate charter to a then unheard-of company, the Shawinigan Water and Power Company, later absorbed in part by Hydro-Québec. The company was empowered to develop water power and produce gas and electricity for light, heat, and motive power. The charter gave it the right to transmit power throughout the province of Quebec and to sell it to municipalities. It also authorized the company to

Between 1870 and 1914, over two million people immigrated to Canada. Many settled in western Canada on land granted to the Canadian Pacific Railway in return for construction of the railway. The Immigration Branch of the Department of the Interior ran a high-powered advertising campaign designed to attract immigrants from abroad. Pamphlets in Swedish, Dutch, and German cleverly suggest that a Utopian version of the traditional homeland culture is available in Canada. Publications such as *Canada West* promoted a healthy, invigorating climate and lucrative wheat yields. The government's campaign continued into the 1920s. (Public Archives Canada; Canadian Pacific Corporate Archives A-12987)

Another attraction of the West was oil, demonstrated by these placards promoting oil company stocks on Calgary's 9th Avenue in 1914. The first oil boom in Alberta prompted intense sales campaigns of low-priced stock by many small companies seeking capital for exploration. Calgary became a speculator's paradise with over 500 companies formed within a year. (Glenbow Archives NA-841-377)

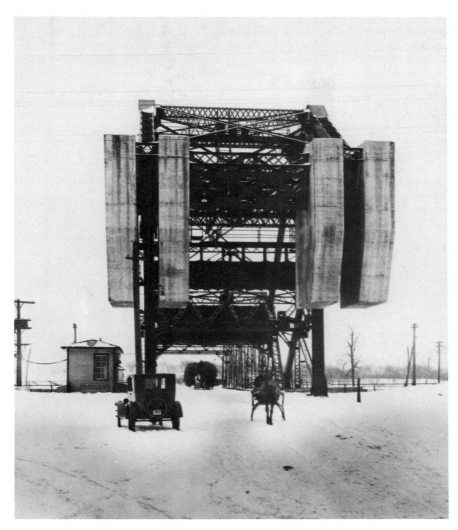

A wide variety of vehicles were used for transportation in early twentieth century Canada. In the photograph above, taken about 1912 on 101st Street in Edmonton, one can see an electric streetcar, a Grand Trunk Pacific steam locomotive, a gasoline-powered Model T Ford, and a wagon drawn by oxen. The photograph to the left is believed to have been taken near Montreal, perhaps on the Lachine Canal. Beneath the massive, concrete counterweights of a canal lift bridge, there are a swift cutter, a sledge heavily laden with hay, and an automobile equipped with chains for better traction on snow and ice. (Canada Cement Lafarge Ltd.; Public Archives Canada C-56695)

The rapid development of western Canada in the early twentieth century is reflected in this photograph of the construction of Edmonton's High Level Bridge in 1911 or 1912. The photograph was taken from the newly finished legislative building. In the foreground are the buildings of an old trading post, soon to be cleared away. (Glenbow Archives NA-1328-1214)

The first Alberta oil boom was ignited when naphtha gas began flowing on May 14, 1914, at a depth of 1158 metres, from the Dingman No. 1 well in Turner Valley. Drilling had started in 1912 after gas was reported bubbling from a large slough in Turner Valley in 1911. This photograph shows a party of visitors to the site in 1914. (Glenbow Archives NA-952-2)

acquire existing works and property, and to expropriate property as needed.

The company already owned an extremely promising power site on the St. Maurice River at Shawinigan Falls. However, Shawinigan Falls was, by the standards of the time, in the midst of the wilderness. The much-heralded, long-distance transmission line from Niagara Falls, New York, to Buffalo had proven the feasibility of distant power transmission, but it covered only 42 kilometres over easy terrain in a long-settled area. By contrast, the Shawinigan Falls plant would have to transmit power 137 kilometres to Montreal over inhospitable terrain in a fiercely cold climate.

This dramatic view looking down the penstocks toward the powerhouse of the Pittsburgh Reduction Company shows the emerging electrochemical complex at Shawinigan Falls. The cables, made of bundles of aluminum rods (in the background), transmitted direct-current electric power from the powerhouse to the smelter and cable plant at the top of the hill. (Alcan Aluminium Ltd.)

One of the original Hall pots was installed at the Pittsburgh Reduction Company, Shawinigan Falls, in 1901 for the electrolytic production of aluminum. Although nothing in the photograph indicates the scale, according to one employee the pots were not much bigger than a bath tub. The Pittsburgh Reduction Company eventually became Aluminum Company of Canada Ltd., more commonly known as Alcan. (Alcan Aluminium Ltd.)

The Niagara Falls development was in the European engineering tradition of facilitating existing patterns of trade, industry, mining, and agriculture. Shawinigan Falls, on the other hand, was in the Canadian pioneering tradition. Like the earlier railroads and canals, it was predicated on the belief that if the structure and the facilities were provided, economic activity would follow.

The early stages of the Shawinigan Falls project were fraught with legal wrangles and financial uncertainty. After the province of Quebec successfully defended its legal claim to the valuable hydro-electric site at Shawinigan against private interests, its ownership was contested by the federal government. In 1898, the British Privy Council settled the dispute in Quebec's favour. This decision, by awarding rights to the beds of navigable and floatable rivers to provincial governments, gave provinces an extremely powerful role in determining the shape of Canadian economic and engineering history.

To ensure its financial viability, the Shawinigan Falls project needed assured sales of its power. On April 29,

1899, the first customer appeared. A group led by former CPR general manager and president Sir William Van Horne, a man grown rich on railroads and other engineering-based investments, agreed to lease 7463 kilowatts of power for the Carbide Company, later known as Canadian Carbide, to produce calcium carbide.

Next, on August 14, 1899, the Pittsburgh Reduction Company signed a long-term contract with Shawinigan to buy 3730 kilowatts of power. The Pittsburgh Reduction Company, through its subsidiary Northern Aluminium Company – later Aluminum Company of Canada Ltd. – produced aluminum by the newly developed electrolytic process which required large amounts of direct current electricity. Because the Shawinigan promoters planned to produce alternating current, the contract was not for generated electric power but for a water supply that would allow the Pittsburgh Reduction Company to generate its own direct current.

In the spring of 1900, the third key customer was signed, the Belgo-Canadian Pulp and Paper Company. Like aluminum, paper production from pulpwood was still a relatively new industry. The company was

To meet a continually expanding demand for electricity, the Shawinigan Water and Power Company began building its second powerhouse in 1911 – seen here in the foreground under construction. The inclined railway to the right of the powerhouse was built to supply construction materials to the site. (Hydro-Québec)

The Shawinigan Water and Power Company relied on the enormous hydro-electric potential of the St. Maurice River Valley. In all, 14 developed hydro-electric sites are located on the stretch of the river which can be seen in this model. (Hydro-Québec)

attracted by the availability of both abundant power and suitable pulpwood at Shawinigan.

The signing of these three customers helped assure the struggling company's survival. Moreover, because customers were establishing industries at the site itself, the town of Shawinigan boomed and created the Shawinigan Electric Light Company, another customer.

By 1901, astute bankers and investors realized that the Shawinigan Water and Power Company was a sound investment. The New York financial firm of Farson and Leach agreed to underwrite a share offering. Capital raised from the sale of shares finally put the company on a firm financial footing.

Construction and engineering activity parallelled the quest for customers and finances. An initial challenge was to build a railroad from the Great Northern Railway line so that supplies could be brought to the site. For planning and construction of the hydro-electric complex itself, experienced engineering talent was drawn from the project at Niagara Falls, New York. Wallace O. Johnson came first as a consultant and within a few months was appointed chief engineer for the project. Engineers from Montreal included members of the consulting firm of T. Pringle & Son, which designed the powerhouse.

The project's long-term viability depended on its ability to transmit power 137 kilometres to the huge market in Montreal. The company signed a potentially lucrative contract for the sale of electricity in Montreal, but the contract required the delivery of power to the city by February 1903, less than two years away.

Working against this tight deadline, Shawinigan engineers were able to use the cold Canadian climate to advantage when they strung temporary power lines across poles fastened to timbers frozen in the winter ice. On February 3, 1903, power from a 50 000-volt line – the highest long-distance voltage line in North America

– surged into Montreal. Capping four years of strenuous pioneering work, it was the longest transmission line in Canada and one of the longest in the world. On March 24, Dominion Cotton Mills, the first users of this power for industrial purposes in Montreal, turned on their switches.

Building on the success of Shawinigan Falls, the Shawinigan Water and Power Company grew into one of the great private power utilities in North America with a worldwide reputation for excellence in engineering and technological achievement. By 1963, when it was nationalized as Hydro-Québec, the company had an installed generating capacity of over 1.5 million kilowatts.

The Shawinigan Water and Power development stands as one of this century's most significant landmarks in Canadian industrial and engineering history. It marks a major step in the growth of hydro-electric and power-systems engineering in Canada.

The Klondike Gold Rush and Northern Expansion

Although it was situated only 137 kilometres northeast of Montreal, the power and associated industrial development at Shawinigan demonstrated the commercial potential of wilderness and frontier engineering in Canada. Nothing did more, however, to direct the world's attention to Canadian frontier areas than did the Klondike Gold Rush of 1898.

The popular image of the Klondike is that of a lone, grizzled prospector collecting a fortune in easily accessible nuggets. The reality is that small groups of miners quickly exhausted the rich placer gold deposits in the rivers, creeks, and nearby low-lying gravels. Hand-dug tunnels and drifts replaced surface works, but these too were quickly worked out. Most of the Klondike's gold was in fact inaccessible to solitary miners.

The remaining gold was to be found in low concentrations in gravel – a great deal of it in valley walls high

Dredges such as the Klondike Number 4 of the Yukon Consolidated Gold Corporation (shown above in rear view in 1914) worked their way up and down creek and river valleys in the Klondike. A chain of buckets at the front of the dredge scooped up sand, gold, and gravel and delivered it to separating equipment inside. As the dredge advanced, tailings spilled out in sinuous patterns leaving the type of landscape shown in this photograph (right) of the Klondike River at the mouth of Bonanza Creek. Dredges had very shallow drafts so entire valleys could be dredged with only minimal damming. (Public Archives Canada PA-103872, C-34243)

above available sources of water. Moreover, much of the gold-bearing gravel was locked in permafrost. If gold mining was to continue in the Klondike, engineers were needed to develop methods of processing large quantities of gravel, of bringing water for mining and processing to high elevations, of supplying power to run machinery, and of melting large quantities of permafrost.

One of the early engineering-based mining innovations in the Klondike was the floating dredge. Floating dredges had been used to recover placer gold and tin in other parts of the world, but never in such extreme cold. Foreign-designed dredges were quickly adapted for operation during all but the coldest weather. In the next few decades, dredges worked their way up the rivers and creeks, literally turning the land upside down. Their endless chains of buckets chewed through the gravel, sent it through separators, kept the gold, and sent the rest out the back end of the dredge where a stacker deposited it in continuous winding rows which are still visible today.

Gravel higher up the valley walls lent itself to hydraulic mining. High-pressure hoses washed the gold-bearing gravel from the cliff faces. The mixture was then sent through sluices to separate gold from gravel.

Hydraulic mining required an abundant supply of water. Many miners depended on an elaborately engineered water supply system that combined open wooden flumes and an inverted syphon made of high-pressure steel pipe. The same installation was used to generate hydro-electric power for lighting and for powering machinery such as dredges and pumps.

Permafrost was one of the most difficult problems facing miners in the Klondike. Direct sunlight melted surface gravel, so that hydraulic miners were able to deal with the problem of permafrost by working their high-pressure hoses across large exposed faces, skim-

ming material off on each pass. For other types of mining, however, the permafrost had to be melted to free the gold from the rock-hard matrix of gold, gravel, and ice. Advances in permafrost thawing techniques allowed increasingly lower grades of gravel to be worked economically.

The decades of continuing mining operation in the Klondike were dependent on large-scale machinery skilfully designed and adapted to northern working conditions, hydro-electric power, and extensive water supply systems. In the Klondike, as in other remote areas, the intimate relationship between engineering and the development of natural resources made possible the extension of Canada's industrial frontier.

Spectacular New Mountain Tunnels for the CPR

The opening up of the prairie wheat lands of Canada's West stimulated much of the country's explosive growth in the early years of the twentieth century. The Canadian Pacific Railway had been built largely to transport agricultural products, and the new farms created from virgin prairie depended on this system to get their produce to market.

As Western farms prospered and traffic grew, the CPR worked assiduously to upgrade lines to make them more efficient, profitable, and safer. The effort to improve the quality of the railroad during the early years of the century led to two bold and spectacular engineering achievements in the mountains of British Columbia – the Spiral Tunnels and the Connaught Tunnel.

Tunnels are useful for avoiding the problems of aboveground construction such as steep grades, punishing curves, frequent trestles and bridges, and the dangers of rockslides, snowslides, and avalanches. But tunnels are almost invariably the more expensive form of roadbed construction. When the CPR was first built,

The Lethbridge Viaduct is one of early twentieth century Canada's most spectacular engineering and construction feats. It was built to allow trains on the Canadian Pacific Railway to cross the Belly River near Lethbridge, Alberta, without encountering steep grades. Begun in 1907 and finished in June 1909 at a cost of $1 334 525, the viaduct was over 1.6 kilometres long and 96 metres high. Both these photographs were taken during construction in 1908. The profile below shows the erection traveller with the riveting traveller behind it. The erection traveller stood 18 metres high, providing sufficient clearance for the passage of loaded flat cars. The assembling cage, suspended from the end of the erection traveller, permitted access to all joints and carried a telephone operator who was constantly in touch with the hoisting engineer. (Glenbow Archives NC-2-299, NC-2-282)

A group of workers and managers watch expectantly as the erection traveller lowers the last girder into place on the Lethbridge Viaduct on June 22, 1909. The concrete abutments had been positioned a year earlier by a different contractor. The ease of fitting the girder into place is a testament to good workmanship and design. Considering film speeds in 1909, it is probable that the two men leaping from the girder were artfully placed in position in the darkroom. Close examination will also reveal one worker standing on another's shoulders on the other end of the girder. (Glenbow Archives NC-2-290)

costs had been kept to a minimum by avoiding tunnel construction wherever possible.

Over the years, however, the steep grades on the CPR west of Kicking Horse Pass in the Rocky Mountains and through Rogers Pass in the Selkirks had proven extremely hazardous, taking a heavy toll in lives and equipment. After over two decades of unsatisfactory temporary construction, the CPR decided that tunnels were necessary.

Spiral tunnels had been used in Europe, but John E. Schwitzer, an Ottawa-born, McGill University engineering graduate, still in his thirties, gave them their only North American application. Schwitzer looped the track at Kicking Horse Pass through two long spiral tunnels.

The upper spiral circled 291 degrees under Cathedral Crags, dropping 16.5 metres in height over a distance of 992 metres. The lower spiral circled 217 degrees beneath Mount Ogden as it dropped 15.5 metres in a tunnel length of 891 metres.

When the Spiral Tunnels opened in Kicking Horse Pass in the fall of 1909, they had doubled the length of track but reduced the grade from 4.5 per cent to an acceptable 2.2 per cent. Whereas it had previously required four engines to haul 644 tonnes through the hazardous pass, now two engines could safely draw some 889 tonnes.

The Connaught Tunnel, the other spectacular tunnel and the longest one in North America, was built through Rogers Pass. Although the grade through Rogers Pass was steep, the Connaught Tunnel was built primarily to solve more serious problems caused by frequent heavy snowfalls and avalanches in the area. Between 1885 and 1911, a great deal of equipment and more than 200 lives had been lost to avalanches.

The Canadian Pacific Railway's Connaught Tunnel, in its day the longest tunnel in North America, saved many lives by protecting trains from the heavy snowfalls and deadly avalanches of Rogers Pass. Here, work proceeds at Portal Number 2, one of the entrances to the tunnel. (Glenbow Archives NA-4598-7)

The Spiral Tunnels looped the Canadian Pacific Railway's main line beneath the Rocky Mountains at Kicking Horse Pass. Although the tunnels were built primarily to reduce steep grades, the wide publicity they received also promoted tourist travel on the railway. (Canadian Pacific 6015)

Conventional engineering methods had been used, appropriately and successfully, to build the Spiral Tunnels, but the innovative techniques employed in the Connaught Tunnel aroused the interest of engineers worldwide.

After an exploratory bore in August 1913, the railway decided to proceed with a straight tunnel over eight kilometres long. Construction began with the drilling of a small, preliminary tunnel not intended for rail traffic. The first tunnel – known as a pioneer bore – was an approximately two metre by three metre cross section, located about 14 metres to the side of the centre line of the projected main tunnel and about three metres above its grade. On December 19, 1915, the two ends of the pioneer bore met with mathematical exactness.

To build the main tunnel, cross-cuts were sent out from the pioneer bore to the main route at intervals ranging from 229 to 305 metres. Headings were then cut in both directions to excavate the main tunnel. Narrow-gauge, compressed-air locomotives within the pioneer bore drew out the rubble. In addition, natural air flow and electric fans in the pioneer bore aided ventilation and reduced the levels of dust and poisonous gases during construction.

Because the innovative pioneer-bore method allowed work on the main tunnel to proceed on a number of fronts simultaneously, the project was completed in record time. Regular trains began using the new, double-track Connaught Tunnel on December 9, 1916, less than a year after completion of the pioneer bore.

The Connaught Tunnel was notable in other respects as well. At 8047 metres from portal to portal, the tunnel was well over one kilometre longer than any other North American tunnel. It lowered the summit elevation by 168.5 metres, reduced the length of the line by almost eight kilometres, and eliminated more than seven kilometres of snowsheds built to protect the line from snowfalls and avalanches. It also eliminated over 2500 degrees – the equivalent of seven complete circles – of power-robbing and problem-creating curves in the line.

Both the Connaught and Spiral Tunnels attracted a great deal of international acclaim for Canadian engineers and contractors, who once again had demonstrated their mastery of the conception, design, and management of major engineering and construction projects.

The Canadian Northern Railway

At the turn of the century, the Canadian Pacific Railway was the nation's sole transcontinental carrier. But the era's spirit of vigorous expansionism and boundless confidence rapidly generated schemes for two new transcontinental railways: the Grand Trunk Pacific Railway and the Canadian Northern Railway.

The plans for the new railroads were put forward in the popular Grand Trunk Pacific National Railway policy of 1903–04 which helped Prime Minister Laurier win a landslide re-election victory. Eventually, however, costs soared, and the railways ran into financial difficulties and had to be taken over by the federal government. Nevertheless, the new railroads, particularly the Canadian Northern, were striking examples of the application of engineering know-how.

Earlier railroads had been built as cheaply as possible in the hopes that further business would allow improvements to be made. But during the construction of the Canadian Northern, a new philosophy emphasizing higher initial investment and more permanent structures gradually emerged.

The Canadian Northern Railway, which ultimately became part of the Canadian National Railway system, started out in 1896 as a smaller railroad, the Lake Manitoba Railway and Canal Company. Construction at first followed the traditional Canadian pattern. Built by the able but cagey contractors, Mackenzie, Mann and Co. Ltd., standards were low. Obsolete and unreliable second-hand equipment ran on light rails laid on narrow, poorly ballasted roadbed. Low freight rates compensated for the frequent breakdowns and unreliable schedules.

Low initial construction standards should not, however, be confused with bad design. Because of poor initial route selection, other railroads often could only be improved by relocating the line. Canadian Northern concentrated on finding routes that could be upgraded without costly relocations. Lines were to have a potential or actual grade no steeper than 0.5 per cent after compensation for curvature.

By 1909–10, finances had improved, and the Canadian Northern began to build to standards unmatched by any other railway: heavier rail, more costly excavation and rock-fill, and fewer temporary trestles.

The new approach marked an important shift in engineering philosophy. Higher quality construction required a greater initial investment. But Canadian Northern engineers believed that lower maintenance and rebuilding costs would make the operation more economical in the long run. Implicit in this approach was a confidence in Canada's future, and a belief that the railways were destined to make an important contribution to that future.

Building a railroad with permanent structures and a consistently easy grade was costly and complex. In British Columbia, 16 bridges and 44 tunnels were required; in 1912, in Ontario alone, 9600 workmen were employed to extend and maintain the Canadian Northern system. In addition to the tracks and rolling stock, Canadian Northern built high-quality auxiliary structures across the country. The Port Arthur coal and ore docks, the works of the Mount Royal Tunnel and Terminal Co., and the Vancouver terminal built on reclaimed swampland are but a few examples of these notable new facilities.

When the Canadian Northern Railway opened for regular traffic in October 1915, it had achieved consistently better grades than any earlier road and was built to allow sustained, high-speed traffic. It was, however, virtually bankrupt and would suffer further financial difficulties.

Despite these problems, the Canadian Northern Railway marked the emergence of a new engineering phi-

Started in 1912, the Brooks Aqueduct was a major landmark in both civil and agricultural engineering. Over three kilometres long and nearly 17 metres tall at its highest, the aqueduct was made of reinforced concrete, a material then regarded with some suspicion. The 19 100 cubic metres of concrete and 27 240 kilograms of reinforcing steel bars used in its construction were brought by rail from Winnipeg. During construction two narrow-gauge railways ran parallel to the aqueduct. One was used for mobile cranes and the other for materials and equipment. After the framework was removed, the reinforced concrete aqueduct's sparse, elegant beauty stood out against the landscape. (E.I.D. Archives and Library, Brooks, Alberta)

losophy of permanence based on faith in the country's future. And by demonstrating the ability of Canadian engineers to build high-quality railway systems, it helped to increase both national and international confidence in the profession.

Irrigation in British Columbia and Alberta

The expansion of the Canadian railway system brought with it enormous industrial and agricultural opportunities, particularly in the West. However, areas such as the Palliser Triangle of the Great Plains and parts of British Columbia's interior suffered from limited and uncertain rainfall. Irrigation was the key to a more diversified and stable economy in these regions.

For large-scale irrigation, an administrative structure that can ensure uniformity of purpose among many competing interests is essential. Canadian engineers, in addition to building the irrigation systems themselves, also played a central role in drafting the necessary legislation.

Recognizing that irrigation would be vital to successful farming in British Columbia and the Northwest Territories – now Alberta, Saskatchewan, and Manitoba – the federal government sent a Canadian-born engineer, John S. Dennis, to examine Australian and U.S. practices and make recommendations for draft legislation. The resulting Canadian Irrigation Act of 1894 followed the Australian pattern which put ownership of the water, hence the right to regulate its use, in the hands of the central government.

The legislation cleared the way for the development of an irrigation system in Alberta in the early 1900s. The federal government provided the necessary surveys, located reservoir sites, and then licensed and inspected the irrigation works. Private capital built the required dams, irrigation channels, and other structures.

The situation in British Columbia was complicated, however, by the fact that water rights for agriculture had been granted to individuals since 1858. These rights threatened to undermine the uniformity of administration needed for large irrigation projects.

The federal Water Clauses Consolidation Act of 1897 took the first step in solving this problem by giving the British Columbia government ownership of all unappropriated water in the province. But the act did not deal with the further problems of assigned but unused water rights, mechanisms for water storage, and the arbitration of disputes. Drawing heavily on the advice of Canadian engineers such as John S. Dennis, the federal government cleared up the most serious of these problems with the British Columbia Water Act of 1909.

Extensive irrigation had begun in British Columbia in the early 1890s when the combined advertising resources of the province and the Canadian Pacific Railway had attracted farmers to the Okanagan Valley in the British Columbia Dry Belt. Thanks in large part to the efforts of engineers in devising administrative structures for large-scale irrigation, the transformation of ranch lands to irrigated commercial orchards and gardens was well advanced by 1910.

In British Columbia, engineers had shown that their specialized knowledge could help lay the legislative framework for irrigation. In Alberta, the legislation was straightforward but the scale of irrigation projects was far larger.

The Canadian Pacific Railway had received over 10 million hectares of land described officially as "fairly fit for settlement"[19] from the federal government. Despite the massive advertising campaigns conducted by the government and the CPR to attract settlers, much of this land remained empty. The CPR found itself holding millions of hectares of semi-arid unsettled land with

little or no prospect of selling it in the foreseeable future.

Irrigation seemed the obvious answer and, when Ontario-born civil engineer William Pearce moved to the West in the mid-1880s, he became an irrigation enthusiast. Irrigation, Pearce felt, could turn the CPR's semi-arid holdings into valuable farmland. He became a good friend of the CPR's William Van Horne and urged him to develop large-scale irrigation projects to enhance the value of CPR lands.

Despite the fact that Van Horne remained unconvinced, Pearce persisted and, along with fellow engineer John S. Dennis, was largely responsible for passage of the Canadian Irrigation Act of 1894.

Finally, in 1903, 20 years after Pearce had seen the potential of irrigating CPR lands, the railroad recognized the validity of his ideas. The CPR exercised its rights to further land grants by acquiring approximately one million hectares of additional land between Calgary and Medicine Hat. The land was an unbroken block which made it well-suited for imposition of the unified administrative system required by large irrigation projects.

John S. Dennis immediately resigned from his federal government job to take charge of the irrigation project for the CPR and, in 1905, William Pearce left the Department of the Interior to join him.

The land grant was divided into three roughly equal sections known as the Western, Central, and Eastern Sections. With its rolling hills and high elevations, the Central Section was less suitable for irrigation, and the various plans were never carried out. However, large-scale irrigation projects were successfully completed in both the Western and Eastern Sections.

The Western Section was finished first. By 1910, over two thirds of the Section had been sold and the irrigation system was in full operation. Water was drawn from the Bow River at Calgary. It then flowed through a 27-kilometre canal where a large earth dam across a natural depression created a reservoir five kilometres long, almost one kilometre wide, and 12 metres deep. Three secondary canals, totalling 406 kilometres in length, ran from the reservoir and connected with 2126 kilometres of distributing ditches.

Work on the Eastern Section began in 1909 with the construction of the Bassano Dam on the Bow River. The dam diverted water to irrigation reservoirs at Lake Newell, Cowoki, Sutherland, and Rolling Hills.

One of the most impressive engineering feats in the Eastern Section was the concrete aqueduct at Brooks. The slim columns, the graceful lines of the water channel, and the sparseness of material made it an aesthetically pleasing as well as a utilitarian structure.

At 6.4 metres wide and 2.4 metres deep, the water channel or flume of the Brooks Aqueduct delivered water at a rate of 18 cubic metres per second. Here, workers take a break during the installation of reinforcing bars in 1914. The flume began to deteriorate as early as 1918, requiring continual repair until it was replaced in 1979 by an earth canal. (E.I.D. Archives and Library, Brooks, Alberta)

The Bassano Dam, begun in 1909 in the area of Alberta known as the Eastern Section of the Canadian Pacific Irrigation Block, was another early Canadian concrete structure. It was designed by CPR engineer Hugh B. Muckleston. In this photograph of June 1, 1912, the concrete spillway is nearing completion while a train transports material across the top. Materials were also positioned by a cableway suspended between the two towers at the ends of the dam. (E.I.D. Archives and Library, Brooks, Alberta, 27-273A,B)

By 1914, the Eastern Section was complete and land went on sale. Before the construction of the irrigation system, the area's great aridity had made it risky even for dry farming techniques. When supplemented by irrigation, however, the Eastern Section's greater sunshine and heat, plus its longer growing season, created near ideal conditions for crops such as alfalfa, corn, potatoes, sugar beets, and fruit. The result was a more diversified agriculture than was possible in many other parts of the province.

Canadian engineers, through their contribution to the drafting of irrigation legislation and their design and construction of the irrigation systems themselves, made a vital contribution to western Canadian agriculture.

The Bassano Dam included 24 individual spillways controlled by sluice gates, each with a span of 8.2 metres. Maximum flow could reach 2830 cubic metres of water per second. In this photograph of September 1, 1911, a reusable form for the spillways is lowered into place on top of steel reinforcing rods. The form is suspended from the cableway which stretched across the dam. (E.I.D. Archives and Library, Brooks, Alberta, A26-106)

In addition, the irrigation structures they built, such as the splendid Brooks Aqueduct and the Bassano Dam, were landmarks in the use of a new and important engineering material – concrete.

The Impact of Concrete: Bridges and Lift Locks
Concrete had been available since early in the nineteenth century, but its widespread use in Canada began only around the turn of the century. Its employment as a construction material made it possible to build structures with more precision, permanence, and sophistication, and its acceptance by the trades and the profession marked an important turning point in Canadian engineering.

A Nova Scotia bridge replacement project marked the first large-scale use of concrete in Canada. Between 1883 and 1887, a million dollars was spent in rebuilding the province's stock of deteriorating wooden bridges. In a paper delivered to the Canadian Society of Civil Engineers in 1888, chief engineer M. Murphy described the introduction of concrete as a substitute for masonry in this reconstruction work.

In the construction of these highway or public road bridges, concrete has borne an important part. It was at first – in 1883 – employed sparingly and with hesitation, but of late it has been used largely and with much confidence. Its use, for the support of the superstructure of iron bridges, was prompted by necessity, because of the scarcity of materials suited for ashlar masonry, the costs of transportation, the want of skilled workmen, and the rapidity with which it could be erected with ordinary labour.[20]

Murphy noted that concrete had been resisted at first because it was felt to be too expensive and unable to withstand the rigours of Nova Scotia's winter. However, he said, the charges had proved unfounded. Concrete piers and abutments were supporting 44 iron bridges with spans ranging from 15 to 49 metres. Other works included two small concrete arch bridges at Cow Bay near Halifax and at Acadia Mines, Londonderry, and retaining walls and dry docks in progress.

Murphy was one of the early converts to concrete, but many of his colleagues were reluctant to use the new material. During the discussion that followed Murphy's paper, C.E. Dodwell explained one of the major reasons Canadian and American engineers used little concrete in comparison with their colleagues in Britain:

If every member of this Society were to be asked the question, "Why is not concrete more extensively used in this country as a substitute for masonry?" Nine replies in ten would probably give two reasons. Firstly, because of its inability to withstand the effects of intense frost, and secondly on account of its excessive cost. The tenth man would perhaps with greater candour reply: "Because I never tried it and do not know anything about it. I have no precedent to go upon, and would be afraid to make the experiment."[21]

In addition to their lack of experience with the material, Canadian engineers resisted concrete because of the poor and uneven quality of the cement available to them. The problem was gradually cleared up in the early years of the twentieth century when high-quality Portland cement became widely available.

Portland cement is a type of hydraulic cement. Hydraulic cements will set and harden under water to form a stone-like mass, and they are generally of superior strength and longevity.

Some hydraulic cements – known as "natural cements" – can be made by burning naturally occurring limestone without any admixture at the burning stage. In Canada, Ruggles Wright, son of Philemon Wright, the founder of Hull, Quebec, began manufacturing a natural hydraulic cement as early as 1830 from agrillaceous magnesian limestone quarried near Ottawa. Others were also active in the field. Pierre Gauvreau, for example, a Quebec-born architect and civil engineer, patented Gauvreau Cement in 1854. Made of stone from Cape Diamond, near Quebec City, it proved very durable in stonework exposed to water and humidity. It was used in three forts built at Lévis, Quebec, around 1870, as well as in other public works projects across Canada.

Some natural cements were excellent. The Niagara-formation agrillaceous limestone near Thorold, Ontario, produced a cement mortar of superb strength which set under water in 10 to 15 minutes. The natural cement from Thorold found its way into many noteworthy early Canadian structures such as the piers of the Victoria Bridge in Montreal.

However, natural cements had one serious drawback: inconsistent quality. The quest for consistently high quality led to the 1824 invention of Portland cement, named after the Isle of Portland, England.

Portland cement is a combination of materials processed before and after firing. When mixed with water and an aggregate such as sand, gravel, or crushed stone, Portland cement binds the materials into a solid stonelike mass, known as concrete. Earlier cements had been used largely as agents to hold brick or stone together. Because of its superior quality, concrete made with Portland cement could be used as the load-bearing agent in construction.

In 1889, the firm of C.B. Wright & Sons of Hull, Quebec, a maker of natural cement, became Canada's first Portland cement manufacturer. In Ontario, a number of small companies rushed into the Portland cement industry in the 1890s. The first plant in the West was opened by the Canadian Pacific Railway near Vancouver in 1893, and others followed around the turn of the century.

Overall, the early years of the Portland cement industry in Canada were characterized by cutthroat competition, instability, and bankruptcies. Finally, in 1909, the Canadian financier Max Aitken, later Lord Beaverbrook, brought stability to the industry by merging a number of companies to form the Canada Cement Company Limited, now Canada Cement Lafarge.

The creation of the Canada Cement Company marked a new stage in the history of Canadian engineering and construction. Inconsistent quality control and improper use had given concrete an uneven reputation in its early years. The newly reorganized Portland cement industry under the leadership of the Canada Cement Company was able to launch an advertising campaign to explain the advantages and the proper use of concrete in construction to engineers and contractors. Given a more knowledgeable construction industry, manufacturers in turn increased their research efforts to produce consistently high-quality and special-purpose cements.

The lift lock at Peterborough on the Trent Canal provides an excellent example of early Canadian leadership in the use of concrete. When it and the iron and steel lock at Kirkfield were opened in 1904, they were the first in North America. There were in fact only three other lift locks in the world – in England, Belgium, and France.

As well as improving water travel on the Trent–Severn Waterway, the lift locks at Peterborough and Kirkfield established a number of important engineering precedents and added to the reputation of the Canadian engineering community.

Conventional mitre-gate locks require large quantities of water each time they are used. In addition, where there is a considerable height to be traversed, multiple mitre-gate locks must be built in flights, and passing a ship through these flights can be extremely time-consuming.

The Peterborough Lift Lock is the largest in the world and provides a good example of early Canadian leadership in the use of concrete. In this 1902 photograph, the Peterborough Lift Lock is under construction. An aerial cableway, suspended from the towers at either side of the site, was used for moving concrete and other construction materials. (Public Archives Canada 6-685)

The lift lock is a simple, almost automatic machine designed to solve these problems. In operation, lift locks are smooth, silent, and swift. Their principles were eloquently described by Walter J. Francis, one of the engineers on the Peterborough lift lock project:

In principle the hydraulic lock may be likened to two immense hydraulic elevators of the simple plunger type, having their presses connected together so that the descent of the one causes the rise of the other. In place of the ordinary elevator platform we have a large water-tight box or tank closed at either end by a gate. The lockage is performed by towing the vessel

Here, workmen remove concrete forms from the central tower of the Peterborough Lift Lock in 1902. As plywood sheets were not yet available, forms were made of planks of irregular widths. This left the striated pattern visible on the finished tower. The then customary lack of safety gear for the workmen high on the tower is also noteworthy. (Public Archives Canada 6-681)

This 1903 photograph of the Peterborough Lift Lock nearing completion shows the structural steel lock chamber supported by a hydraulic ram. The designers embellished the concrete towers and approaches with fine architectural details such as cornices and broken belt courses. (Public Archives Canada 16-826)

The steamer Islinda leaves the lower chamber of the Peterborough Lift Lock in about 1915. Each chamber of the lift lock has a gate at either end for letting boats in and out. When the gates are closed, water is pumped into the upper chamber to make it slightly heavier than the lower one. As the upper chamber falls, it forces the lower chamber up. (Public Archives Canada)

into this box of water and then closing the gate on the end of the box as well as that of the canal, thus leaving the box independent of the reach and free to move vertically. The box with the water and the floating vessel is then raised or lowered to the other reach. The chamber or box about to descend is loaded with a few inches more water than the other chamber, thus giving it the necessary additional load or "surcharge" to enable it to cause the ascent of the other when water communication is established between the two presses.[22]

The lift locks at Peterborough and Kirkfield were monumental in scale – the biggest in the world. The largest lift locks in Europe had chambers 42.7 metres long and 5.8 metres wide. Water depth was 2.4 metres. The movable lock chambers at Peterborough and Kirkfield were the same length and their normal depth of slightly more than 2.4 metres was not significantly greater than that of their European counterparts. However, the clear width of 10 metres in the Canadian locks gave them about double the water load.

Moreover, while the 15.2-metre lift at Kirkfield was marginally higher than the highest European locks, the Peterborough lock, at 19.8 metres, or almost seven storeys, towered above all others.

The 19.8-metre stroke and the 2.3-metre diameter of the massive rams at the Peterborough lift lock were also unprecedented. It is believed that they were the largest hydraulic presses in the world. In addition, the presses were made of cast steel, rather than the traditional cast iron. Fabricated by John Bertram & Sons Tool Works of Dundas, Ontario, the cast-steel parts were subjected to thorough testing which provided important empirical data on the characteristics and high-pressure performance of this material.

Yet another design innovation in the Peterborough Lift Lock was the construction material. The superstructure was made entirely of concrete, rather than the conventional stone and mortar. Superintendent of the Trent Canal, Richard B. Rogers, a local engineer who designed the locks, had been largely responsible for the decision to build lift locks in the first place, and was also a strong advocate of the use of concrete. The towers and associated concrete work on the Peterborough lock represented the world's largest single use of massed unreinforced concrete. The equivalent structures at Kirkfield were steel.

The original design, use of materials, and construction of these locks were a monumental tribute to Canadian engineering and construction skills. Like the earlier work of John By and John MacTaggart on the Rideau Canal, the lift locks on the Trent Canal made the best use of available technology and resources. But by the time the lift locks were built, Canadian engineering had entered a new era of sophistication.

New materials such as concrete and advanced techniques such as the use of hydraulic rams demanded a higher degree of precision in engineering than would have been possible in earlier eras. Commenting on the crucial importance of precision during the design and construction of the Peterborough Lift Lock, engineer Walter J. Francis said, "This fact, more than all others, probably, was impressed upon the author during the progress of the work – Eternal Vigilance is the Price of Accuracy, and Accuracy is the Price of Successful Operation in a Hydraulic Lock."[23]

The lift locks also required a higher initial investment than conventional mitre-gate locks. But Richard B. Rogers had successfully demonstrated that lift locks would be less expensive to maintain and significantly more efficient. Rogers had calculated that a fleet of 12 barges and one tug could pass through the lift locks in under two hours, while passage through conven-

Both the Kirkfield and the Peterborough lift locks are part of the Trent–Severn Waterway. The Kirkfield lock operates using the same hydraulic principles, but the towers supporting the lock chambers were built of structural steel instead of concrete. The Kirkfield Lift Lock is seen here on opening day July 6, 1907. Both lift locks are still in operation today. (Public Archives Canada)

tional locks would require more than eight. Like the Canadian Northern Railway, the lift locks embodied the new Canadian engineering philosophy of permanence and quality.

Across Canada, many engineers played a part in the increasing acceptance of concrete. One of the most influential of these was Frank Barber, an extremely talented Canadian engineer and a master of the early twentieth-century concrete bridge.

Barber's work epitomized his belief that the best engineering required an increasingly sophisticated scientific and mathematical foundation. But Barber was also well-rounded and literate; he had been editor of the

Varsity, the University of Toronto student newspaper. As an engineer he wrote extensively, propounding the view that beauty, scientific precision, and cost-cutting were all important and compatible goals in engineering. He chastised his colleagues who "believed that an ugly bridge can be made cheaper than a beautiful bridge. Now in my judgment this is all wrong. Nature never wastes material in accomplishing her results. God never made anything ugly in making it swift or strong."[24]

The bridge that Barber called "the darling of his heart"[25] was located, like the lift locks, at Peterborough. On the fiftieth anniversary of the founding of the Engineering Institute of Canada, C.R. Young described

Still in use today, Frank Barber's beautiful Peterborough Bridge is famous for the daring use of concrete in its low, graceful arch. The lower photograph shows another part of the bridge under construction. The appearance of a double arch results from the removal of the form work after the concrete has set. (Phyllis Rose)

The Osborne Street Bridge, built over the Assiniboine River in 1912, greatly facilitated the growth of south Winnipeg. Bridges which crossed water barriers – the Bloor Viaduct over the Don River in Toronto is another example – opened up new land for development and contributed significantly to urban growth. The awkward concrete superstructures on the Osborne Street Bridge concealed counterbalances that raised the bridge for river traffic. They served as rather ostentatious city gates until they were removed in 1937. (Manitoba Archives)

Barber's Peterborough Bridge as "one of the outstanding concrete bridges of Canada for structural and architectural merit."[26] The choice of this bridge as the only modern illustration in the article on arches in the eleventh edition of the *Encyclopedia Britannica* is further testimony to Barber's design genius.

The Peterborough Bridge was technically innovative – representing the first use of temporary hinges in Canada – and daring in the use of materials. Its centre arch of 71.6 metres was for years the longest span of its kind in Canada. Designing with mathematical precision, Barber insisted that concrete forms follow the theoretical lines as closely as possible; he would not resort to the easier-to-make, but heavier, cruder shapes.

Barber consulted architect Claude Bragdon who suggested design changes that called for less concrete and made the bridge more streamlined. Bragdon also added a handrail which emphasized the structure's long

The increased use of motor cars and concrete revolutionized transportation in the early twentieth century. This 1912 photograph shows a steam powered mixer discharging concrete into a horse-drawn carrier. The concrete was used to pave St. Mary's Road in St. Vital (near Winnipeg). (Manitoba Archives)

In 1921, the Canada Cement Company began construction of its innovative 10-story head office building in Montreal. Cement products were incorporated into the building wherever possible to demonstrate the material's advantages for both structural and decorative purposes. Rather than the more customary steel frame construction, the floors, load bearing members, and the roof were made of reinforced concrete. Slump and compression testing was used; a precise aggregate to cement ratio improved the strength of the concrete. The exterior facing was poured with an exposed aggregate to give the more luxurious appearance of stone. The building, which was also one of the first in Montreal to have an underground parking garage, is still in use today as the headquarters of Canada Cement Lafarge Limited. (Canada Cement Lafarge Ltd.)

The Banff Springs Hotel, a resort hotel located in the Rockies, is one of Canada's most familiar landmarks. Its facing of stone and brick gives the hotel a traditional chateau-style appearance. Actually, the latest in reinforced concrete technology was used when the hotel was built in 1926. In this photograph, the south wing is nearing completion. The marks left by the wooden planks used for concrete forms are clearly visible. (Canada Cement Lafarge Ltd.)

horizontal lines. The final result was a bridge whose economy and beauty of line were a visual delight.

Barber worked on many other remarkable bridges – the Leaside Bridge in Toronto is a good example. By 1929, he could write that "altogether I have been in responsible charge of the design and supervision of over four hundred bridges."[27] Today, Frank Barber's concrete bridges stand as a testament to one of the great pioneers in the development of concrete as an aesthetically appealing, modern building material.

As the century progressed, engineers used concrete in an increasing number of structures, including drydocks, roads, harbour works, hydro-electric dams, public buildings, sewers, factories, mills, and agricultural buildings of all sorts – perhaps most characteristically in grain elevators.

The dramatic upsurge in wheat production necessitated a rapid increase in storage capacity. In harbours

The automobile and the road improvements it necessitated transformed Canadian cities. In this photograph, taken in September 1928, workmen convert the intersection of Côte d'Abraham and St. Rial in Quebec City to a concrete roadway. (Canada Cement Lafarge Ltd.)

While the history of Canadian engineering is characterized by its outstanding accomplishments, there have also been a number of failures. The wooden grain elevators that dotted the prairies were generally satisfactory. On occasion, however, they did collapse. In 1911, for example, the Laura Elevator near Saskatoon collapsed, spilling its contents across the railway tracks and blocking them. (The British Library 24792)

and railyards where land was expensive, cylindrical reinforced concrete grain elevators offered the dual advantage of considerable height and maximum internal volume in relation to surface area. As a result, concrete elevators proliferated at grain terminals throughout North America.

The increasing use of concrete in early twentieth-century Canada was both the result and the cause of a trend toward increasing permanence, quality, and precision in construction. Progress was not always smooth. In 1913, for example, the Transcona concrete grain elevator, loaded unevenly but according to traditional techniques, listed about 30 degrees. A structure of any other material would have been demolished, but concrete was so strong that it could be righted. This amazing feat by a young engineering and construction company, the Foundation Company of Canada, served to increase confidence in concrete. During the early years of the twentieth century, improvements in the quality of concrete, along with increasing confidence and familiarity with the material, made the transition to this new type of construction possible and changed forever the face of Canadian engineering.

The Quebec Bridge: Triumph over Disaster

Accounts of the stirring triumphs of Canadian engineering in the early part of this century also call to mind the occasions in which engineers have had to overcome crises and failures. One of the most serious of these –

The enormous one million bushel, concrete elevator in Transcona, Manitoba, remained intact when, in 1913, uneven loading caused it to list about 30 degrees. After the grain was removed, engineers of the Foundation Company of Canada excavated and poured a new foundation. The building was then jacked back into position. (Foundation Group)

a double disaster – was the collapse of the Quebec Bridge.

On the morning of August 29, 1907, Canadian bridge erectors were pushing an engineering form to new limits. The Quebec Bridge was to be the longest cantilever span in the world and the pride of the Phoenixville Bridge Company of Phoenixville, Pennsylvania. Later in the day, it became a sickening tangle of structural steel which enclosed the bodies of 75 workmen.

After a prolonged investigation, the contract for a new bridge was signed on April 4, 1911. The contractor was the St. Lawrence Bridge Company, a new firm created by the Dominion Bridge Company of Lachine, Quebec, and the Canadian Bridge Company of Walkerville, Ontario.

The St. Lawrence Bridge Company began work immediately and construction proceeded outward from both shores of the St. Lawrence River. On September 11, 1916, the centre span which had been floated into the river started its slow ascent into position. It never arrived; it plummetted into the river, this time killing 13 men. Once again the Quebec Bridge seemed doomed. Fortunately, the cantilevers were not damaged. The following year, a new suspended span was floated into the river and successfully connected into final position on September 21, 1917.

A Royal Commission determined that the tragic collapse of the first Quebec Bridge in 1907 was caused by faulty design and inadequate supervision. This photograph of the main pier is from the collection of the Royal Commission. (Public Archives Canada C-9766)

The Quebec Bridge was a magnificent accomplishment. Much of the work had been completed under wartime conditions and all of it under the shadow of earlier failure. Its 548.6-metre clear cantilever span was the longest in the world – 30.5 metres longer than the renowned Firth of Forth Bridge. In addition, it could carry two and a quarter times the live load of the Firth of Forth Bridge.

Structurally, the successful Quebec Bridge incorporated the first major use in a bridge of the "K" system of web bracing. Conventional cantilever bridge con-

Completed in 1917, the Quebec Bridge was the longest cantilevered bridge in the world. The innovative "K" system of web bracing is clearly visible in the photograph above. The girders were made at the St. Lawrence Bridge Company in Montreal, formed expressly to carry out work on the project. In the lower photograph, taken in 1916, a wartime guard protecting the bridge against sabotage is dwarfed by the colossal size of the structure. (Public Archives Canada PA-56401, PA-135835)

struction employed the uncertain procedure of leaving compression joints open and unriveted until the increasing dead load on the bridge closed the joints. Proposed by Phelps Johnson, president of the St. Lawrence Bridge Company, the "K" system avoided the necessity of leaving joints open during construction.

The September 27, 1917, issue of the prestigious *Engineering News-Record* detailed the many contributions of the Quebec Bridge to engineering:

> *Before closing the final chapter in the design and erection of this remarkable structure, it is proper to record the debt that bridge-builders owe to the work at Quebec. It has advanced greatly our knowledge of the problems of large compression members and of tension bars. The effects of distortion in trusses were explored farther than before and means devised for dealing with such effects. Much knowledge has been added to our store of experience on the assembly of heavy members, while new standards were set as to degree of precision and finish in shopwork. Then there is, beyond all this, the great gain in our general grasp of the problem of very large bridges as to practicability and cost.* [28]

The *Engineering News-Record* went on to describe the bridge's contribution to engineering as a profession:

> *To the individual engineer the great value of the achievement lies in the inspiration emanating from the courage of the men who have erected on the failure of 1907 and the loss of 1916 this greatest of bridges – and in so doing not only have erected a monument to themselves and their courage and ability, but have vindicated the profession before a doubting world.* [29]

There was, however, little time to celebrate. Global warfare was continuing to occupy public attention, as the entire country mobilized in an allied effort to defeat the German and Austro-Hungarian forces in Europe.

Building and Manning the War Machines

As the accompanying photographs show, the resource and heavy industries were geared up specifically for a gigantic war effort, thrusting unprecedented challenges at the engineering profession.

These were challenges that were to be posed with even greater force a little more than two decades later, as Canada, which suffered immense losses in the First World War and emerged a different – and in many ways stronger – nation as a result, was plunged again into a fearsome global war.

Rien à faire sans l'emprunt de la victoire (nothing doing without victory bonds) declares this 1918 poster. It is a reminder that, without the money raised through victory bonds, factories would be frozen and machinery remain idle. (Canadian War Museum, National Museums Canada, J-20020-3)

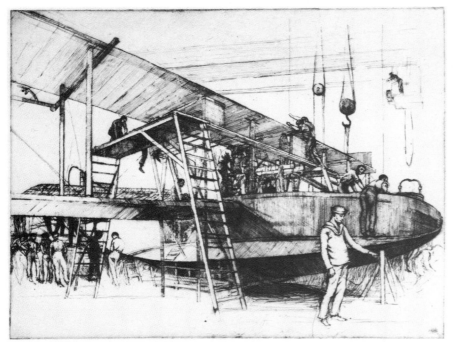

During World War I, the Imperial Munitions Board placed many orders with Canadian shipyards. A total of 134 vessels comprising 20 wooden schooners, 69 wooden steamers, and 45 steel steamers were built by British Columbia shipyards. This photograph, taken in about 1917, catches the launch of a freighter from J. Coughlan and Sons shipyards in Vancouver. (City Archives, Vancouver)

In April 1918, Canadian Aeroplanes Limited in Toronto, operated by the Imperial Munitions Board, received a contract from the United States Navy for 30 twin-engine F5 flying machines. With a wingspan of 31 metres, the F5 flying machines were at that time the largest airplanes ever built in North America. This 1918 etching was done by Dorothy Stevens, an artist commissioned to record the war effort. (Canadian War Museum, National Museums Canada, 8826)

All across Canada, factories retooled for wartime production. Desperate shortages of workers resulted in a heavy influx of women in positions formerly held almost exclusively by men. In this 1917 photograph, women munitions workers at Vancouver Engineering Works Limited prepare shell casings for shipment. (Vancouver Public Library 910)

During World War I, many factories were converted to wartime production. But the British Forgings Limited plant at Ashbridges Bay in Toronto was built with great speed in 1917 specifically for wartime service. The caissons for the melting house foundations were in place by March 1, 1917; by July, the plant was nearly ready for operation. The melting house's 10 Heroult electric furnaces can be seen above. Below, a worker in the melting house's smoky interior is protected only by heavy clothing and dark glasses. (City of Toronto Archives SC-125-67, SC-125-74)

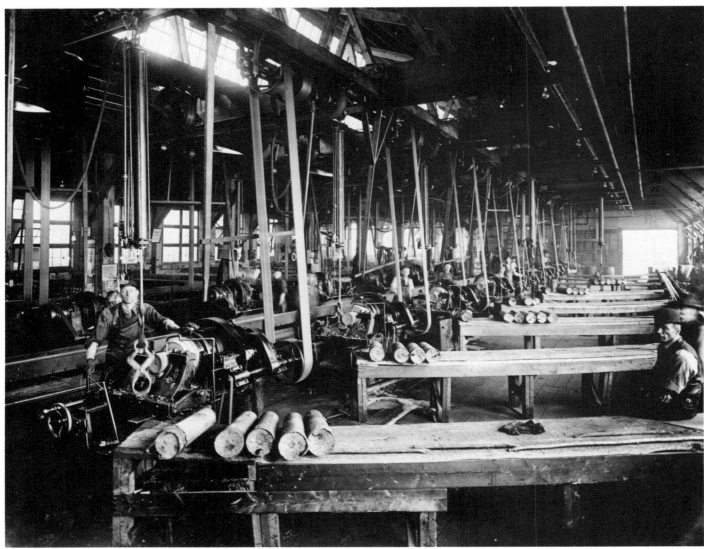

The cutting and breaking shop was another section of the British Forgings Limited plant built in Toronto for wartime production. The absence of protective eye wear and the unguarded belts of the overhead power transmission system show a lack of regard for worker safety which was typical during this period. (City of Toronto Archives SC-125-77)

4

Decades of Hope and Depression: The 1920s and the 1930s

The popular image of the years between the the two World Wars is one of stark contrasts: the decadence and conspicuous prosperity of the Roaring Twenties, followed by the despair and poverty of the Dirty Thirties.

However, the interwar years were far less uniform than these stereotypes suggest. The early years of the 1920s in Canada were, in fact, a time of significant economic hardship. There were a record number of business failures each year from 1921 through 1923; during the winter of 1921–22, half of the school children in a low-income suburb of Vancouver exhibited symptoms of malnutrition.

Furthermore, although the Depression brought an end to the economic expansion of the latter half of the 1920s and was catastrophic for many, for others it brought almost no change. Major areas of the economy and therefore of engineering – mining is a prime example – experienced an almost constant increase in activity throughout the 1920s and 1930s. Overall, through both economic prosperity and hardship, engineering became continuously more important in the lives of Canadians during the interwar years.

This can be seen throughout a broad range of industrial, social, and economic activities during this period – for example, the expanding mining and pulp and paper industries; the mass marketing of the motor vehicle, with the subsequent road building programs; the increase in urban planning; the invention of the snowmobile; the growth of public works during the Depression; and the emergence of industrial engineering.

Each of these posed new challenges to the Canadian engineering profession, and each challenge, and its outcome, left an imprint on Canadian society.

Rise of Mechanized Mining

The development of agriculture in the West had been the critical driving force behind the expansion and prosperity of the opening two decades of this century. Now, however, attention focussed on the resources of the North and, in particular, on mining.

There was widespread public enthusiasm for mining. The Precambrian Shield became a topic of everyday conversation; promoters talked about a northern treasure box bursting with wealth; and it was commonly believed that Canada was poised to enter the greatest mining era in history.

Production figures supported the rhetoric. Between 1921 and 1929, silver production doubled and gold trebled; nickel, lead, and zinc quadrupled; copper rose sevenfold. The workforce for these six non-ferrous metals increased from 10 000 to 23 000 during the same period. The value of production reached $150 million a year in 1929 and, with the exception of a few bad years, continued to climb throughout the Depression.

The mining boom opened up new parts of the country. Traditionally, mining had been concentrated in Ontario and British Columbia. Now, the major new mining and smelting complexes in Rouyn Noranda, Quebec, and Flin Flon, Manitoba, stimulated development in these provinces as well.

New technologies and new engineering techniques boosted the demand for minerals, and the increased supply of minerals, in turn, fed the burgeoning industries. Lead, copper, and zinc were consumed domestically and exported, particularly to the large United States market, to support the new and expanding automobile, electrical, and radio industries.

Virtually every aspect of the booming mining sector depended heavily on engineering. For example, mechanized transport was making prospecting significantly more effective. With an outboard motor, prospectors could cover far more territory during the ice-free season than they had been able to do in cedar strip canoes. The airplane increased their effectiveness even more,

Canada's machine tool industry played an important role in supporting the hydro-electric, railway, and mining industries. In this industrial equivalent of the class photograph, proud machinists and workers at John Bertram and Sons of Dundas, Ontario, pose on the 11-metre boring and turning mill they assembled for Canadian Westinghouse Limited of Hamilton. After the mill was assembled and tested in Dundas, it was disassembled, transported to Hamilton, and erected on the foundations shown in the photograph below. The mill is still used today in manufacturing equipment for mines and hydro-electric generating stations. (Public Archives Canada C-111729, C-111726)

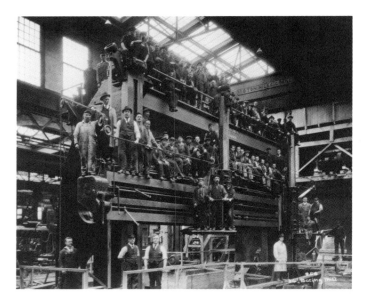

allowing them to cover in a few hours or a single day an area that might have previously taken months. Equipped with skis in winter and pontoons during the rest of the year, the airplane opened up year-round prospecting and made it possible to transport sophisticated heavy equipment directly and swiftly to the sites.

Technology and engineering were also changing the mines themselves. In the Klondike, the small holdings of individual owner-operators had been replaced by company-owned dredges working low-grade gravels in big leaseholds. Large-scale mechanized mining came to predominate in other areas of the country as well.

The mines of the Canadian Shield are a good example. Canadian Shield rock is among the hardest in the world, mines are located far from major population centres, and, by earlier standards, many of the ores would normally be considered low grade. However, Canadian engineers developed solutions to these problems and, as a result, the hardrock mines of the Canadian Shield were to become largely responsible for the reputation Canada acquired as a leading mining nation.

One of the important contributions that engineers made to hardrock mining was the perfecting of the compressed-air core drill that was used to extract core

Compressed air drills are being used in this photograph of the Arntfield Mine in Quebec taken during the 1920s. Modern equipment such as this was behind the dramatic early twentieth century expansion of mining that made Canada famous for its mines and mining engineers. (Public Archives Canada C-17033)

Canadian railroads fostered the growth of a vast engineering and industrial support system. This lathe was manufactured by John Bertram and Sons for turning locomotive drive wheels as large as 2.5 metres in diameter. (Public Archives Canada C-111741)

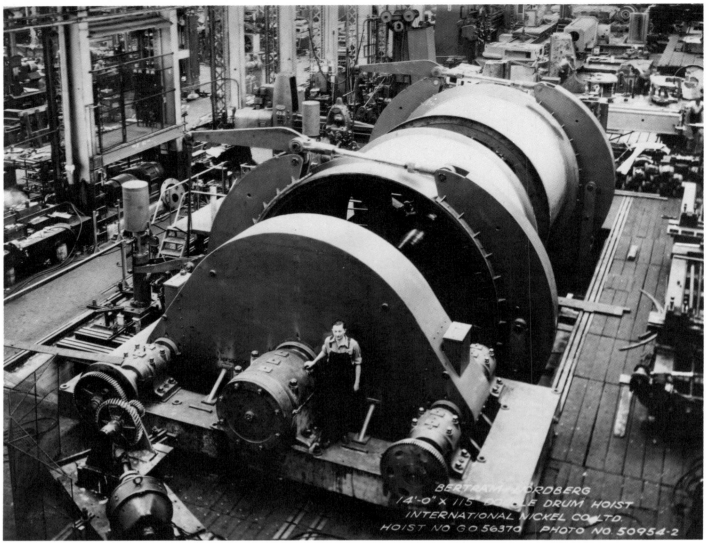

BERTRAM NORDBERG
14'-0" X 115 DOUBLE DRUM HOIST
INTERNATIONAL NICKEL CO. LTD.
HOIST NO G0 56370   PHOTO NO 50954-2

Mining, like the railroads, stimulated significant growth in support industries. Here, a Bertram/Nordberg 4.3-metre diameter double drum hoist is shown ready for delivery to International Nickel. (Public Archives Canada C-111712)

samples. These samples were analyzed for their ore content and the results, carefully plotted on maps, determined where and how miners would drill, blast, and remove the ore.

Another contribution was the production of specialized machines for haulage, both vertically and horizontally. Steam or electrically driven drum hoists, electric underground trams, and automatic-loading, self-dumping skips all helped to move ore farther and faster and increased the amount of ore a mine could handle.

Other innovations included better methods of stoping, improved electric lighting and signalling in the mine, more efficient pumps, and the development of solid rock drills and high-powered explosives. In addition, advances in chemical and metallurgical knowledge and equipment allowed lower grade ores to be worked.

The result was that mines were no longer producers of a limited volume of high-grade rock. Instead, they became high-volume producers of both rich and low-grade ores. Rather than carefully following a single rich vein of ore, miners could work large underground areas that contained both high and low grades of ore.

The Hollinger gold mine in Timmins, Ontario, epitomized the new approach. With a staff of 3000 working around the clock, Hollinger was the second largest gold mine in the world. Its systematic orderliness led one observer to note that gold appeared to be manufactured rather than mined.

The below-ground mining methods pioneered in the Precambrian Shield were matched by similar above-ground innovations in milling, separating, smelting, and refining.

The Sullivan Mine in the East Kootenay region of British Columbia is an apt example. During the First World War, the Consolidated Mining and Smelting Company had begun an electrolytic zinc recovery plant to supply crucial wartime needs from the ores of the Sullivan Mine. The same ores contained lead, but this was difficult to recover.

In August 1920, a differential flotation process was first used to concentrate the lead. The process was successful and regional lead production doubled within three years. A 1923 Department of Mines publication hailed the region's production as a feat "achieved by the economic mining of an immense tonnage at the Sullivan mine and the successful application of the most up-to-date metallurgical practice."[30]

Differential flotation was only one part of a complex aboveground lead mining process at the Sullivan Mine. The process depended on modern engineering in vir-

tually every respect. Heavy haulage equipment was designed to handle large quantities of ore. Mucking machines were introduced to replace expensive labour. Because differential flotation worked only if the ore was ground very finely, a hydro-electric plant was built to supply the large amounts of power required.

During the 1920s and 1930s, then, mechanized mining was replacing hand tool mining. The new processes depended on the development of the machinery and systems that made it possible and profitable to find, extract, and process large quantities of relatively low-grade ore. This was the special contribution of the mining engineer.

The complex machinery needed by engineering-based mining was expensive, however, and corpora-

Specialized equipment for refining ore was an important contribution of engineers to hardrock mining. This 1936 photograph shows the Denver Fahrenwald flotation machines at the McIntyre Mine in Schumacher, Ontario. Machines such as these consumed huge amounts of electric power and generally required large hydro-electric facilities near the mine. Fortunately, rich mineral deposits and excellent hydro-electric sites were both relatively plentiful in the rugged Precambrian Shield. (Public Archives Canada PA-17575)

tions were required to finance such capital-intensive ventures. Gradually, the mining industry became the domain of large companies.

As a result of these developments, mining became far more sophisticated, both in technical know-how and operational management. Mining engineers were expected to maximize the profits of large corporations and also to deal with a broad range of technical issues in metallurgical, chemical, mechanical, electrical, and civil engineering.

Increased public attention on mining during the inter-war years brought the names of famous mines such as Flin Flon, Hollinger, Porcupine, Dome, McIntyre, Rouyn, Cobalt, Timmins, and Trail to the fore, and their exploration and marketing successes seized the popular imagination.

Canadian mines and mills were important sources of vital materials, munitions, and supplies during the First World War. Pictured here is the zinc tank room in the electrolytic zinc recovery plant of the Consolidated Mining and Smelting Company, Trail, British Columbia. The plant was built to support the war effort. (Public Archives Canada PA-15521)

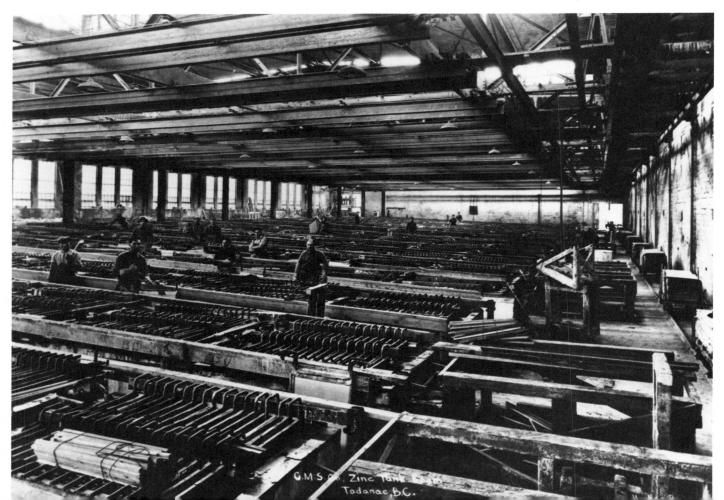

The 1920s were marked by rapid growth in the Canadian pulp and paper industry. This photograph, taken in about 1920, shows the grinding room in the Anglo-Newfoundland Development Company pulp and paper mill at Grand Falls, Newfoundland. Pulp and paper mills were a major source of challenging employment for engineers. As well as supervising daily activities, they designed and built the mills, machinery, access roads, hydro-electric dams, and generating stations. (Public Archives Canada PA-135109)

This increased attention and the growing demands on mining engineers gave enhanced responsibility and status to the profession. Much was being demanded of it and, as the record shows, much that was both practical and theoretical was being delivered.

The Pulp and Paper Boom

Mining may have captured the public imagination during the interwar years, but the pulp and paper industry had far greater economic impact.

Progress in the industry was slow but steady throughout the nineteenth century and up to the end of the First World War. Although there is some controversy regarding the exact dates, it appears that Canada's first paper mill had started operation as early as 1803; the first wood-pulp mill was built in 1866, and the first sulphite mill in 1885.

During the 1920s, however, the pulp and paper industry grew at an astonishing pace. Newsprint production for 1919 stood at 654 000 tonnes, rose to 1.3 million tonnes five years later, and hit 2.5 million tonnes in 1929.

By 1929, there were 108 mills in six Canadian provinces; almost half of the production capacity was in Quebec, a third in Ontario, a tenth in British Columbia, and the remainder in New Brunswick, Nova Scotia, and Manitoba. As a result, 33 000 workers were directly employed in the pulp and paper industries and another 180 000 were partially dependent on it for their employment.

On the demand side, foreign markets were the key to growth. About 90 per cent of the newsprint was exported, some 75 per cent to the United States.

In the United States, the Underwood Tariff Act of 1913 had opened the market to Canadian exports. U.S. supplies of pulpwood were fast dwindling or inaccessible, and the burgeoning U.S. newspaper industry needed reliable sources of economically priced newsprint. The demand for newsprint reflected the needs of a growing and better educated population. American per capita consumption trebled between 1900 and 1925 and continued to grow during the 1920s as the Sunday newspaper grew thicker and more popular.

An additional fillip to growth in the demand for newsprint was the action of the governments of Ontario, Quebec, British Columbia, and New Brunswick, which forbade the export of unprocessed pulpwood derived from Crown Lands. This restriction ensured that the high value-added end of pulp and paper processing took place in Canada, rather than in the United States.

Expansion on the supply side of the industry combined a number of factors, including railway construction which had opened previously inaccessible areas, the growth of hydro-electric generating capacity which provided the plentiful supplies of power and water needed, the availability of investment capital, and

The pulp and paper boom stimulated the demand for hydro-electric power and the construction of facilities such as the Seven Sisters Falls Power Plant and Dam. Located about 90 kilometres northeast of Winnipeg, Manitoba, Seven Sisters was built by the North Western Power Co. Ltd., an associate company of the Winnipeg Electric Company. It began operation in 1930 and had a full capacity of 225 000 horsepower. (Canada Cement Lafarge Ltd.)

the activities in all of these related areas of trained engineers.

In the 1920s, pulp and paper emerged as a major engineering activity. Sawmilling during an earlier age of timber exploitation had required neither first class permanent construction nor sophisticated machinery, and power requirements were relatively low. As a result, there was generally little demand for engineering expertise.

Pulp and paper mills, by contrast, were complex chemical and manufacturing installations requiring a high level of engineering. Concrete buildings carefully built to accommodate complex chemical processes and flow paths replaced rough timber sawmills. Hydro-electric plants were built to meet expanded power requirements.

Canadian engineers played a key role in developing equipment and facilities to meet the demands of the rapidly expanding market. The February 1927 issue of the *Pulp and Paper Magazine*[31] described a number of Canadian engineering innovations that had improved the productivity and profitability of the industry. These included insulation to reduce heat loss, better filters, systems that improved the recovery of reagents and the use of waste products, and more long-lasting machinery and buildings.

Canadian engineers who made signal contributions to the industry worldwide include John Seaman Bates and George H. Tomlinson. Sr. Bates, who had received a doctorate in chemical engineering from Columbia University, developed the process for the clarification of kraft green liquor at Bathurst, Nova Scotia, in 1922. The process saved considerable money and reduced a vexing pollution problem. Tomlinson developed an improved method for the recovery of heat from burning pulp liquor.

Canadian designed and manufactured machinery for the pulp and paper industry was particularly important in the 1920s boom. Dominion Engineering Works Limited produced the first two Canadian newsprint machines in 1920. All such machines had previously been imported. The new machines set a world record for high-speed newsprint production and launched Dominion Engineering and its subsidiary, Charles Walmsley & Company (Canada) Limited, on a period of dramatic expansion.

By the early 1930s, the two companies had built 41 newsprint machines, nine paper-making machines for book and specialty papers, and two pulp-drying machines. By the mid 1930s, about 50 per cent of the newsprint made in Canada was produced on machines made by Dominion and Walmsley.

The companies' machines incorporated a number of important innovations, including anti-friction bearings, an enclosed dryer drive running in oil, and the grouping of the dryer drive in sections to reduce felt and paper stresses. The new suction couches and suction presses were essential for the safe running of high-speed newsprint machines.

The pulp and paper mills themselves were remarkable feats of construction engineering. Many consulting and contracting engineers were involved, but the work of Morrow and Beatty Limited, construction engineers from Peterborough, Ontario, provides a good example.

Born in Peterborough in 1866, Harold Archibald Morrow had been involved in canal and bridge construction in Ontario and the Maritimes after his graduation in 1887 from the Royal Military College in Kingston. He went into partnership with James A. Beatty of Toronto in 1908, and thereafter worked in the new engineering areas of hydro-electric power development and pulp and paper mills in northern Ontario, Quebec, and New Brunswick.

In 1915, Morrow and Beatty completed the Iroquois Falls pulp and paper mill for the Abitibi Power and Paper Company. With a daily production of 499 tonnes of newsprint, the mill was one of the largest in North America. The huge complex had been made of the new materials, concrete and steel, throughout, and construction had taken 18 months.

The hydro-electric system at Iroquois Falls included an overflow dam about 213 metres long. The 122-metre-long powerhouse contained 16 hydraulic units, each generating 1194 kilowatts under a head of 13.7 metres. Twelve of the 16 hydraulic units were vertical and were connected to grinders for making ground pulpwood. The remaining four were horizontal units operating electric generators.

After Iroquois Falls, Morrow and Beatty continued their impressive record in hydro-electric and pulp and paper plant construction. These include the Smooth Rock Falls pulp mill and hydro-electric development of the Mattagami Pulp and Paper Co.; the Drummondville hydro-electric power plant for Southern Canada Power Co. Ltd.; the Grand Falls hydro-electric development in Bathurst, New Brunswick, for the Bathurst Company Limited; the harnessing of the Twin Falls in the Abitibi River for the Abitibi Power and Paper Company; and the construction in 1933 of a hydro-electric generating plant on the Kapuskasing River for the Spruce Falls Power and Paper Company.

As in mining, sophisticated engineering in the pulp and paper industry required high levels of investment capital, and ownership became concentrated in the hands of large corporations. By 1930, three companies – International Paper, Abitibi Power and Paper, and Canada Power and Paper – controlled 55 per cent of the industry, and 81 per cent was controlled by the top six companies.

During the early 1930s, the effects of the Depression were being felt by the pulp and paper industry. New construction stopped and an industry-wide shakeout put renewed emphasis on engineering to ensure the productive and competitive use of resources and facilities. New orders for equipment dried up, for example, and Dominion Engineering was called upon to replace worn parts of its newsprint machines with improved components as part of an industry-wide retrofitting program. After 1934, demand and production began to climb again.

For the two decades of the twenties and thirties, Canadian engineers had played a central role in developing a modern pulp and paper industry capable of profitably and productively meeting the demands of a rapidly expanding market. Pulp and paper mills and their associated hydro-electric power developments were massive and complex engineering projects. They represented the work of large numbers of engineers before, during, and after construction, and helped contribute to the sense of accomplishment and spirit of cooperation that existed in the profession through buoyant periods of expansion, subsequent cutbacks, retooling, and recovery.

Cars, Trucks, and Roads
Given Canada's vast distances, difficult terrain, and extremes of weather, it is not surprising that Canadians took to the automobile enthusiastically. By the 1920s, Canada was second only to the United States in the number of automobiles per capita. Motor vehicle ownership soared from 408 000 in 1920 to well over 1.2 million a decade later.

Most of the automobiles were built in Canada. A hefty 35 per cent ad valorem duty was levied on imported vehicles and foreign car companies were quick to establish branch plants in Canada. The result was that, among secondary manufacturing industries, only the automotive industry could match the growth of primary sector industries such as mining and pulp and paper during the 1920s. The 68 408 passenger cars built in Canada in 1919 climbed to 188 721 by 1929.

Most Canadian companies in the industry were custom coach builders producing a limited number of fine automobiles for the wealthy. The few companies involved in mass production provided the greatest number of automobiles and tended to be U.S.-owned.

Ultimately, mass production came to dominate the market and the custom manufacturers went out of business. The result was that automotive engineering jobs moved south of the border. However, there were many other ways in which the automobile industry created opportunities for engineers.

The most direct and immediate consequence of the growing popularity of the automobile was the need to upgrade existing roads and bridges. In theory, the motorcar was meant to adapt to pre-existing roadways made for horse-drawn vehicles. In practice, it required surfaces that would not become impassible quagmires with spring thaw or rain, and bridges that would not collapse under heavier loads. Determining how best to meet local and regional needs required a considerable amount of engineering research. Of the numerous agencies pursuing this goal, it appears that some of the most innovative work in understanding actual road-building conditions was sponsored or conducted by the Ministry of Highways, established by the province of Quebec in 1914.

Canada had over 482 800 kilometres of roads in 1920, but less than 1600 kilometres were paved. Many city, county, and consulting engineers began laying out and supervising the resurfacing with concrete and asphalt of existing dirt and gravel roads. Bridges were upgraded or replaced to carry heavier and wider truckloads.

During the 1920s, traffic volume multiplied dramatically. Trucks, in particular, were rapidly increasing in size and number as they serviced the growing residential, commercial, and industrial developments in suburban areas.

Gradually, planners realized that passenger cars, trucks, and buses needed roads specifically designed for them, rather than upgraded horse-drawn vehicle roads or new roads modelled after older ones. One of the principal tasks facing highway engineers was improvement of the alignment, visibility, and grades of main arteries so that traffic could move safely and quickly.

The problem was particularly acute in southern Ontario. By 1930, the Ontario Department of Highways had developed an extensive network of highways to serve most of the region, but in densely populated areas, traffic congestion threatened to overwhelm the system.

A new highway between Hamilton and Toronto had been recommended as early as 1916. But when construction finally started in 1929 on what was called the Middle Road, progress was slow. With a change of government and a new Minister of Highways, the Middle Road was redesigned to become the Queen Elizabeth Highway, and construction was rapidly completed between 1937 and 1939.

The Queen Elizabeth Highway marked Canada's entry into the new age of road construction. When it opened in 1939, the Queen Elizabeth was Canada's first divided highway and one of only a handful in North America. It was a major engineering accomplishment and a symbol of the growing commitment to find new ways to accommodate the automobile.

The Queen Elizabeth was designed to provide rapid transit and a high level of safety. One of the keys to

The artist's depiction above shows John Lyle's original conception of the northwest entrance to Hamilton near the Queen Elizabeth Way. In Lyle's proposal, the High Level Bridge, and the Desjardins Canal cut beneath it, served as efficient transportation arteries. The area was also intended to be a centre for boating. The photograph below of the completed highway shows that Lyle's original design was considerably altered. Nevertheless, the decorative approaches above the bridge abutments preserved a sense of his classic style. (Ontario Archives, John Lyle Collection)

safety was the divided highway. Although U.S. parkways were divided, none was as long as the Queen Elizabeth, at 144.8 kilometres.

Controlled access was used to improve both safety and traffic volume. The number of intersections crossing the road was minimized and access from adjoining private property was restricted. In Grimsby, underpasses were built to route the highway under existing streets. Canada's first cloverleaf, partial cloverleafs, and traffic circles also reduced the number of intersections.

The aims of the Queen Elizabeth Highway's planners were to promote tourism, particularly from the United States, and to attract industry to what is now known as the Golden Horseshoe, the curve of lakeshore between Toronto and Niagara Falls.

Tourists were to be enticed by a beautiful road as well as fine vacation sites. Architects, landscape architects, town planners, and engineers were all involved in the highway's design and construction. When it was completed, it was widely praised as a beautiful and scenic route. In a study of the highway, historian John C. Van Nostrand has placed it in the context of a movement aimed at making public works both beautiful and efficient.[32]

The Queen Elizabeth Way, as it became known, has been so successful in attracting industry that nearby settlement and traffic volume have grown continuously. Since 1947, the road has undergone numerous modifications to handle the increasing traffic. The design of these alterations has tended to be strictly utilitarian, with little regard for the original intention of creating a highway that was to be attractive as well as functional. Nevertheless, the Queen Elizabeth Highway stands as a landmark at the beginning of the modern era in Canadian highway engineering and construction.

Planning for Urban Growth
The rapid increase in the number of cars and trucks in Canada did more than speed the construction of superhighways; it radically altered traditional patterns of living and working as well. Motorized vehicles allowed people to live farther from their place of work, and industry was able to move from urban centres into outlying areas with access to motor vehicles. The result was rapid suburban growth which had started with electric street railways.

Inside cities, road systems were not designed for the burgeoning numbers of motorized vehicles and traffic congestion became a perennial problem. At the same time, many cities were facing the deleterious effects of inadequate sewers, outmoded water systems, and substandard housing.

During the 1920s and 1930s considerable thought and energy were expended on making Canadian cities more livable, efficient, and beautiful. These goals were not, however, always viewed as compatible and there was a good deal of acrimonious debate over urban planning. Two slogans, the City Beautiful and the City Efficient, expressed the major issue in its starkest terms. There was a tendency, although far from an invariable one, for engineers to champion efficiency and architects, beauty.

Marius Dufresne, municipal engineer for the city of Maisonneuve from 1910 to 1918, was a notable exception to this general rule. Dufresne, an engineer who could double as an architect, demonstrated a concern for both beauty and efficiency.

Advertised as the Pittsburgh of Canada, Maisonneuve, now part of Montreal, had experienced extremely rapid urban growth based on industry and engineering. With excellent rail and water links, and adequate supplies of power and labour, Maisonneuve enjoyed the country's fifth highest industrial output. However, the city was

All across Canada, growing cities demanded better transportation facilities. The Bloor Viaduct over the Don River in Toronto, shown here under construction in July 1917, was an ambitious public works project many years in the planning. It opened up the Danforth area for development and eventually carried a subway line. (City of Toronto Archives 841)

Shortly after the Toronto Transit Commission took over a private firm, the Toronto Railway Company, on September 1, 1921, it rebuilt the Queen and Church Street intersection. Here, two Toronto Transit Commission workmen are perched atop a truck-mounted mobile work platform while their supervisor looks on from below. The forest of poles in the background is a reminder that, as late as 1921, many utilities were still privately owned. During the early decades of electric service, it was common to find many companies competing to supply service to a single area. Each company would erect its own poles. (Toronto Transit Commission Trans-234)

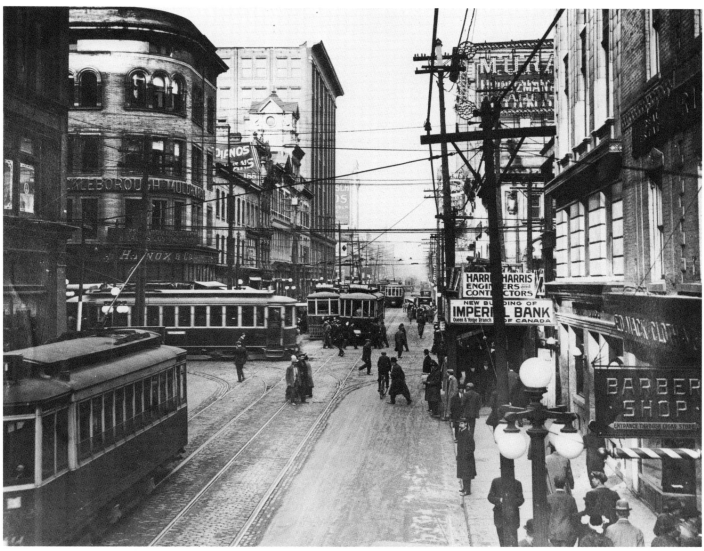

Toronto Railway Company streetcars crowd the intersection at Yonge and Queen streets in 1915, as pedestrians dodge to avoid them. The Toronto Railway Company cars were built in the company's own shops at Front and Frederick streets in Toronto. The company lost almost one third of its fleet in two disastrous carhouse fires during the First World War. (Toronto Transit Commission)

Although hydro-electric installations were built primarily to meet commercial and industrial demand, residential customers provided additional market opportunities. Across Canada, electric companies launched campaigns to inform housewives about the benefits of electricity. These 1929 photographs show a mobile kitchen sponsored by the North Shore Power Company in Quebec. (Hydro Québec)

also inflicted with the rather crowded ugliness and blandness characteristic of so many rapidly growing industrial centres.

Marius Dufresne was extremely well-informed about historical and contemporary planning, architecture, and engineering in North America and Europe. Inspired by the Parks and Boulevard Movement in Europe, he was one of the moving forces behind the beautification schemes that bore considerable fruit before being cut short by World War I.

Unlike many of his reform-minded engineering colleagues, Dufresne was supported by a group of wealthy businessmen with political power. Dufresne beautified Maisonneuve by building large boulevards and parks that led to and were flanked by imposing public and private buildings. Four important public buildings were designed and built within a period of five years.

The classically inspired Town Hall of 1912 was followed two years later by an indoor Public Market, complete with ultramodern facilities and a political meeting hall. The magnificent Public Bath and Gymnasium, which opened in 1916, was modelled on the lines of the recently completed Grand Central Station in New York City. Similarly, the modern-looking Fire Station, which opened in 1915, shows familiarity with the work of Frank Lloyd Wright, one of the freshest and most innovative architects of the day. Chateau Dufresne, the private residence built by Marius and his brother Oscar, is another component of the Dufresne legacy. Modelled on the architecture of the Petit Trianon at Versailles, it is now a museum of the decorative arts.

Marius Dufresne's role in the beautification of Maisonneuve, his ideas on transportation facilities for the Island of Montreal, and subsequent work with his own engineering company demonstrate that Canadian engineers were not intellectually isolated and that they were keenly aware of, benefitted from, and contributed

Electrification radically changed the appearance of the urban landscape. These two photographs, taken during the 1920s at the corner of Vitré and Saint-Urbain in Montreal, illustrate the impact of electric wires and poles. (Archives National du Québec, Montreal, Fonds Conrad Poirier)

to international developments. More detailed studies of the work of individual engineers such as Dufresne are much needed.

Canadian engineers adopted a variety of means to help improve the urban environment. In London, Ontario, for example, an engineer wrote a series of newspaper articles promoting the construction of a better flood control system for the Thames River. The articles were persuasive and the system was built.

Winnipeg provides another excellent example of the improvements engineers were able to bring to the quality of urban life in Canada. In the early years of the century, Winnipeg was one of the fastest-growing municipalities in North America. Regarded as the new Chicago of the Canadian West, Winnipeg, like Chicago, suffered from the problems of inadequate urban facilities, systems, and equipment. The lack of an adequate supply of potable water led to frequent outbreaks of diseases such as typhoid fever.

Opened in 1919, the Winnipeg Aqueduct was both a major factor in improving public health and an important work of engineering. The aqueduct supplied the city with 363 680 litres of fresh water a day from Shoal Lake, 156 kilometres away. Health conditions improved dramatically during the 1920s and Winnipeg lost its reputation as one of the continent's unhealthiest cities.

In 1924, Winnipeg was responsible for another innovation in urban engineering. The Amy Street Steam Plant of Winnipeg Hydro began heating part of the city's downtown core with surplus steam from its 11 190-kilowatt (15 000-horsepower), coal-fired, thermal-electric generating station. A cogeneration plant for electric power and centralized heating was extremely advanced for the time.

The Winnipeg Aqueduct and the Amy Street Steam Plant marked the beginning of Winnipeg's long tradition of bold and imaginative public works engineering.

The Lions Gate Bridge across the First
Narrows in Vancouver was a toll bridge
built by Guinness brewing interests to provide access to their luxury subdivision, the
British Pacific Properties. Property development was delayed by the Depression and
the Second World War after the bridge
opened in 1938. After the war, however,
the well-travelled bridge became a widely
recognized symbol of Vancouver and British
Columbia. (Vancouver Public Library 9631)

Urban growth and concentration fuelled a
demand for large office towers in city cores.
In this photograph, taken during the late
1920s, workmen carefully position carved
stonework on the Canada Life Building –
known for its flashing weather beacon – at
University and Queen in Toronto. Growing
commuter traffic is reflected in the numerous parking lots in the background below
on York Street. The workmen's felt or wool
caps, soft-toed boots, and lack of safety
harnesses would not be tolerated today.
(City of Toronto Archives 116)

The Winnipeg Aqueduct stands as a monument to excellence in concrete construction and engineering. The project was particularly notable for its mechanized mixing, handling, and placing of concrete. Reusable form work, much of it moved by travellers on rails, and rail-mounted, mobile concrete mixing plants allowed construction to proceed very quickly. (Canada Cement Lafarge Ltd.)

J. Armand Bombardier was one of Canada's most successful entrepreneur-inventors. In Bombardier's earliest vehicles, the problems of transportation under severe winter conditions were solved by ingenious adaptations of existing technology. In 1928, he modified a Model T Ford, replacing the front wheels with skis and attaching double wheels covered with a steel track at the rear. (Musée J. Armand Bombardier)

Other cities followed Winnipeg's example, and the improvement of city services such as water, sewers, and electric power were part of the dramatic growth in urban engineering activity which began the radical transformation of Canadian cities in the twentieth century.

Eric Stevens and Art Hobson, high-school students from Saskatoon, were typical of the many Canadians who experimented with home-built snow vehicles during the Depression. (Public Archives Canada C-129853)

Bombardier and the Snowmobile

In a country plagued by heavy snows, motorized snow vehicles were a natural product of the widespread tinkering and experimentation with the new technology of the internal combustion engine. Today, because of the success of Bombardier Inc., we tend to equate J. Armand Bombardier exclusively with mechanized snow vehicles. In fact, all across Canada numerous individuals tried to use engines to fight the isolation and inconvenience of winter.

The work during the 1930s of two Saskatoon high school students, Eric Stevens and Art Hobson, was typical. Using parts that were either homemade or scrounged from junkyards, Stevens and Hobson created a series of gasoline-powered, propeller-driven snowmobiles. The vehicles were built primarily for fun and the boys and their friends raced up and down the frozen South Saskatchewan River at breakneck speeds. Local doctors also used the snowmobiles for house calls. Unlike Bombardier, however, neither Hobson nor Stevens thought of the commercial application of their invention and they eventually abandoned it.

The skidoo revolutionized winter transportation in activities as diverse as farming, trapping, and racing. The heart of the Ski-Doo was the drive system developed by Armand Bombardier. (Bombardier Inc.)

The early work of J. Armand Bombardier was indistinguishable from that of the numerous other snowmobile inventors. But for Bombardier, the development of a practical vehicle that would reduce the often dangerous isolation imposed on Canadians by winter snows became a lifetime mission.

The key technical element that made Bombardier's snowmobiles superior to others was the continuous belt drive which ran on the snow. When told by suppliers that his patented belt drive could not be manufactured, Bombardier solved all the problems and made it himself, building, in the process, one of Canada's largest engineering and manufacturing conglomerates.

Public Works during the Depression

With the 1929 crash of the stock market and recurring crop failures in the West, Canada found itself locked in the Depression. Factories closed and that symbol of prosperity, the automobile, was transformed by many farmers into the "Bennett buggy," an automobile pulled by the horses it had once proudly replaced, and named after R.B. Bennett, the Tory Prime Minister who had promised relief from hard times.

Bombardier's essential contribution to snow vehicles was the sprocket wheel and track drive system which he invented in 1935 and patented in 1937. In this 1936 photograph, a rubber-covered metal drive sprocket engages a continuous one-piece rubber belt equipped with cross-links. The sprocket drives the belt past three wheels – two large and one small – which are linked mechanically. The system was first used on Bombardier's seven-passenger snow vehicle, the B7. (Musée J. Armand Bombardier)

The Bennett Buggy, an automobile drawn by a horse, was named after Prime Minister R.B. Bennett and was a common sight during the Depression. Here, a Bennett Buggy passes in front of the University of Saskatchewan. (University of Saskatchewan Archives)

Engineers did not escape the problems of the Depression and many found themselves without jobs or working for reduced pay. But much that happened during this period highlighted the essential role engineers could play in helping the country work its way through straitened economic circumstances.

All across Canada, public agencies looked to labour-intensive public works projects to provide employment for the long lines of men on relief. Before the Depression, construction project planners had tended toward a greater use of machinery and manufactured components to reduce labour costs. The new conditions called for a different approach to design and construction – one that was well illustrated in the Depression-era work of C.J. Mackenzie.

After graduating with a degree in civil engineering from Dalhousie University in Halifax, C.J. Mackenzie moved to Saskatoon in 1910. Varied and productive work in a private consulting practice in Alberta and Saskatchewan led him to a teaching position at the University of Saskatchewan.

In the early 1930s, Saskatoon turned to road, sewer, and water system construction, public buildings, and the paving of lanes and sidewalks as relief projects. One of these projects was the Broadway Bridge designed by C.J. Mackenzie.

The bridge was to link Nineteenth Street and Broadway Avenue across the South Saskatchewan River. A steel bridge was rejected because it would be too expensive and employ too few people. Instead, Mackenzie designed a beautiful, multiple-arch, concrete structure that has become a Saskatoon landmark.

The Broadway Bridge was built largely by unskilled local labour. As many as 450 men were employed at one time. Work proceeded around the clock in three shifts and in weather as cold as minus 40 degrees celsius. Despite the use of unskilled labour, a high level of workmanship was maintained. Mackenzie also added to an already impressive reputation for innovative concrete construction – his work on alkaline-resistant concretes was well known – by establishing new standards for building with concrete at low temperatures.

C.J. Mackenzie was consulting engineer on another important relief project as well, a labour-intensive concrete bridge over the North Saskatchewan River at Ceepee, 50 kilometres outside Saskatoon on the highway to Edmonton. The three-span, bowstring arched bridge was actually designed as a thesis project by B.A. Evans, one of Mackenzie's students at the University of Saskatchewan.

The importance of the labour-intensive design of the bridge was summarized by historian David Neufeld:

Skilled labour for the erection of the framework and the operation of limited equipment on site came largely from Saskatoon. However, the bulk of the labour force consisted of local farmers. They hauled sand and gravel with their teams through the heat of the summer and in one of the coldest winters on

This 1933 photograph of the Broadway Bridge in Saskatoon shortly after its completion was taken by the bridge's distinguished designer, Dr. Chalmers J. Mackenzie. As a Depression relief project, the bridge was designed to maximize the use of local labour and materials in its construction. The Bessborough Hotel is visible through one of the arches of the bridge. (University of Saskatchewan Archives)

These two photographs show the progress
of construction on the Broadway Bridge in
Saskatoon between April 20 and August 10,
1932. The Depression era bridge, which is
featured on the dust jacket of this book, is
still in use. (Canada Cement Lafarge Ltd.)

record moved over 13 000 tons [11 791 tonnes] of concrete in wheelbarrows.[33]

As consulting engineer for the bridge at Ceepee, Mackenzie directed design and supervisory work on the site to his students, providing them with practical experience in an era of poor job prospects. The close relationship between the University of Saskatchewan's College of Engineering and relief work during the Depression is part of a long tradition of useful community-oriented research that began with the school's founding in 1912 and continues in the College of Engineering to this day.

Mackenzie's work in Saskatchewan illustrates the important part engineers played in helping the country survive the ravages of crop failures and financial collapse during the Depression.

The Glenmore Reservoir Dam in Calgary was begun in 1929, just before the Depression. However, during the Depression, the dam was a major source of relief work. This photograph emphasizes the graceful, almost sculptural lines of the dam's concrete arches and spillways. The beautiful lines are the by-product of a design meant to minimize erosion and damage from fast-flowing waters. (Canada Cement Lafarge Ltd.)

All over the hard-hit prairies – the Palliser Triangle was fast becoming a desert – engineers were active, under the federally funded Prairie Farm Rehabilitation Administration, in devising methods to reduce soil erosion, and in developing catchment basins, irrigation systems, and stock-watering dams. This conservation work has continued, and today the agency still operates programs on engineering services and rural water development.

Emergence of Industrial Engineering

Between the many businesses that closed their doors during the Depression and the few that prospered stood the huge majority that struggled on, surviving despite plummetting sales and profits. For many of these firms, survival required greater efficiency – an area in which engineers had already made substantial contributions and in which they would come to play an increasingly important role in the years ahead.

It was Frederick Winslow Taylor, an American born in 1856, who, along with some of his key disciples, set the tone for much early twentieth-century management and industrial engineering. "Taylorism" represented the gospel of efficiency through scientific management techniques such as time-and-motion studies.

Taylorism, or what in Canada became known as the York Plan, had one of its early successes in Canada in 1927, when Charles Bedaux and his team of efficiency experts were called in to improve productivity at the York Knitting Mills in Toronto. Bedaux and his colleagues worked with a young foreman and chemical engineer at the plant, Ralph Presgrave. After two years, they devised a system, the York Plan, which successfully increased productivity.

Presgrave, with Bedaux's training, was now able to go out on his own. He applied the York Plan success-fully to the Hamilton plant of York Knitting Mills. Then, with the blessing of J. Douglas Woods, president of York Knitting, Presgrave and two of his colleagues, A.W. Baillie and E.D. MacPhee, formed the industrial engineering firm known as J.D. Woods & Co. and later, after a merger with the accounting firm of Clarkson Gordon, as Woods Gordon.

At first, most of their work was done for York Knitting Mills, but other clients were soon found. One of their early successes was the manufacturer of shock absorbers, Canadian Acme Screw and Gear. In 1935, although the company was making a good product it was not making a profit. J.D. Woods & Co. turned the business around within a few months. As a result, Canadian Acme Screw and Gear was able to reduce labour costs and turn a profit – even in the middle of the Depression.

By applying the precise analytical skills that they had learned as engineers to industrial management, Presgrave and his colleagues were able to help companies weather the Depression. They joined the ranks of other engineers who, throughout the interwar years, made a wide range of vital and lasting contributions to the country.

5

A Swift Response to the Demands of War

The Second World War – 1939 to 1945 – spurred the most massive mobilization of manpower, resources, and industry in history and precipitated an end to the Depression.

With most of Europe overrun or under siege, Canada contributed much-needed personnel, supplies, and equipment to the immense, all-out war effort mounted by the Allied powers against the Nazi-Fascist Axis. From the shop floor to the Cabinet – where engineer and politician C.D. Howe, as Minister of Munitions and Supply, was responsible for organizing Canada's wartime production – engineers played a pivotal role.

During the Second World War, military power became increasingly dependent on science and technology. Vital wartime research – including work on the atomic bomb, radar, and cold-weather operation of aircraft – was carried out under the auspices of the National Research Council of Canada. C.J. Mackenzie, onetime Dean of Engineering at the University of Saskatchewan, was appointed to head NRC in 1939 and to orchestrate the activities of Canadian scientists and engineers in strategic wartime research and development.

The single most important contribution of Canadian engineering to the war effort, however, was in supplying vast quantities of manufactured goods and essential materials – airplanes, ships, weapons, aluminum, steel, plastics, and synthetic rubber, to name just a few.

Power and Aluminum for Aircraft Production

The Canadian aircraft industry provides perhaps the best illustration of the massive mobilization of engineering expertise, resources, and industrial capacity during the war years. The number and variety of airplanes that Canada produced required the organization of engineering and industrial capacity on a large scale. It also depended on a chain of engineering-based activities leading from the design and construction of hydro-electric power projects to the opening of new mines and aluminum plants.

Aircraft Production

After the war, J. De N. Kennedy, wartime Deputy Minister of the Department of Munitions and Supply, summed up the accomplishments of the department's Aircraft Production Branch: "One of the most dramatic achievements of Canadian wartime industry was the transformation of the country's small, relatively unimportant peacetime manufacture of aircraft into large scale production of fighters, bombers and trainers."[34]

Growth in the aircraft industry was spectacular. At the beginning of the war, Canada produced about 40 airplanes a year. The industry employed about 4000 people in eight small plants with a total floor space of some 46 500 square metres. By the end of the war, production had increased a hundredfold to 4000 airplanes a year, and the industry had 116 000 employees on the rolls and 1.4 million square metres of floor space.

Canada assumed complete responsibility for manufacturing all of the training aircraft required by the British Commonwealth Air Training Plan. In addition to producing relatively unsophisticated trainers, Canada was a major manufacturer of such advanced warplanes as the Lancaster bomber. A four-motored, long-range aircraft, the Lancaster maintained excellent manoeuvrability while carrying the largest and heaviest bomb loads. Victory Aircraft Limited, a Crown company located in Malton, Ontario, achieved a peak production rate of 50 Lancasters a month.

A number of private Canadian companies also converted to airplane production during the war. The Canadian Car & Foundry Co. Limited combined forces with Fairchild Aircraft Limited to manufacture Curtiss Hell

As Minister of Munitions and Supply during the Second World War, C.D. Howe was one of the most powerful individuals in Canada. His appointment was due in large part to his reputation and experience as an engineer. C.D. Howe and Co. Engineers were particularly famous for their rapid construction of grain elevators. These two photographs, taken only four months apart, show the fast construction by Howe's company of the United Grain Elevator in Port Arthur (now Thunder Bay), Ontario, in 1927. (Canada Cement Lafarge Ltd.)

During the Second World War, Victory Aircraft, a Canadian Crown corporation, produced as many as 50 Lancaster bombers a month. This photograph commemorates the fortieth Lancaster bomber completed at Victory's Malton, Ontario, plant. Sixty-five wartime women workers stand on the huge plane's wings. (Maclean's Magazine)

The Hawker Hurricane, produced by Canadian Car & Foundry Co. in Fort William, Ontario, was regarded as one of military aviation's most outstanding and versatile planes. A day and night fighter, it played an important role throughout the war, most notably in the Battle of Britain. (National Aviation Museum 6036)

Divers, one of the most powerful dive bombers in the war. Canadian Car & Foundry also produced as many as 106 Hawker Hurricane fighters a month. Elsie MacGill, who had been the University of Toronto's first woman engineering graduate in 1927, was chief aeronautical engineer at Canadian Car & Foundry during the war.

The history of Canadian engineering contains the names of some exceptional women, but perhaps none as exceptional as Elsie MacGill. Although unusual, she did what many other Canadian engineering graduates had to do: emigrate to the United States to find challenging professional work. In Michigan, her employer, the Austin Automobile Company, entered the aeronautical industry and so did Elsie MacGill. In 1929, she became the first woman to graduate from the aeronautical engineering program at the University of Michigan. By 1934, MacGill was an aeronautical engineer with Fairchild Aircraft Limited in Longueuil. Four years later, she became chief aeronautical engineer for Canadian Car & Foundry Co. Here, she designed the Maple Leaf Trainer and, during the war, managed the production of Hawker Hurricane fighters. She was in charge of as many as 4500 workers who produced the Hawker Hurricane at the company's plant in Thunder Bay (then Fort William), Ontario. Wartime production of these planes totalled 2000 including the winterized version equipped with skis and de-icers designed by MacGill.

Much of the credit for Canada's exceptional record of aircraft production belongs to Clarence Decatur Howe, Minister of Munitions and Supply. Arguably the most

These photographs show Hawker Hurricanes being built at Canadian Car & Foundry's plant in Fort William, Ontario. Although it was designed as a land-based fighter, the Hawker Hurricane could also be equipped with bombs or fly from and land on aircraft carriers. On catapult ships and merchantmen, the Hurricane was launched from deck-mounted catapults. Because these ships had no landing facilities, pilots had to make for land after their missions. (National Aviation Museum 4200, 17965)

Fairchild Aircraft Limited of Longueuil, Quebec, reached a peak production of 123 Curtiss Hell Divers a month during the Second World War. Designs for the plane originated outside the country, but 60 000 modifications were incorporated during the war while production lines continued to roll. (National Aviation Museum 6804, 6802)

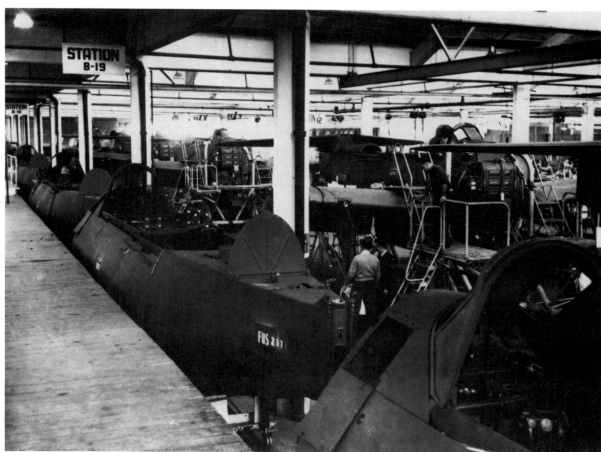

influential engineer in Canadian history, C.D. Howe was a superb organizer and administrator. He had an uncanny ability to analyze problems, find the right people, and motivate them to strive for solutions.

The management of Canada's wartime production of over 16 000 airplanes was a challenging task, combining demanding schedules with the need to master complex designs and manufacturing techniques. Many of the original designs came from outside the country. However, design flaws had to be corrected and continual changes made to reflect new developments in technology and in aerial warfare techniques and to cope with chronic wartime shortages of materials. In the Curtiss Hell Diver alone, 60 000 modifications had to be incorporated while production lines kept rolling.

J. De N. Kennedy, Howe's Deputy Minister, described the management job as "comparable to the manufacture of a gigantic jigsaw puzzle comprising thousands of parts, accessories and pieces of equipment, all of which had to combine with such exact tolerances that a wing-tip for a Lancaster could be built in a Canadian factory and flown to Gander, Newfoundland, or to an operating base in Europe or India, with the certainty that it would fit to a thousandth part of an inch."[35]

Aluminum

Large-scale aircraft production required vast quantities of aluminum. Anticipating the outbreak of war, the Aluminium Company of Canada Ltd. – Alcan – decided in 1938 to build a huge new plant capable of converting aluminum into the forms required by the aircraft industry.

Design work for the new plant started on March 1, 1939, and the first working plans were ready by March 15. Property on the outskirts of Kingston was purchased on May 1, and ground was broken three weeks later. The first concrete was poured on June 1 and the first structural steel went up in the middle of July.

War broke out before the steelwork could be completed and, as a result, necessary pieces of equipment were in short supply. Nevertheless, on July 9, 1940, only 13 months after breaking ground, the first extrusions were produced. By the end of August, the hot mills were rolling.

The Kingston plant became a critical aluminum war production facility supplying millions of kilograms of strong alloy sheets, tubes, forgings, and extrusions for aircraft builders in Canada, Britain, and the United States. It operated on a continuous 24-hour-day, seven-day-week cycle. At its peak, it employed 3750 people.

The prodigious output of the plant was part of an extended supply chain feeding the wartime aircraft industry. Kingston needed aluminum, which in turn depended on bauxite and electric power.

Bauxite

Guyana was the major source of bauxite for the Allied war cause. The task of organizing bauxite mining to meet wartime needs fell to Leslie Parsons, a McGill University civil engineering graduate. When he arrived in Guyana in September 1938, Parsons had just finished building the Bayer process aluminum plant at Arvida (now Jonquière), Quebec.

Drawing on the traditional strengths of Canadian engineering in large and complex construction projects, Parsons was able to increase Guyanese bauxite production by more than five times. The massive undertaking involved opening new mines, building a diesel electric railway to the site, dredging the Demerara River to handle large ships, and increasing the capacity of crushing, washing, drying, storage, and loading facilities.

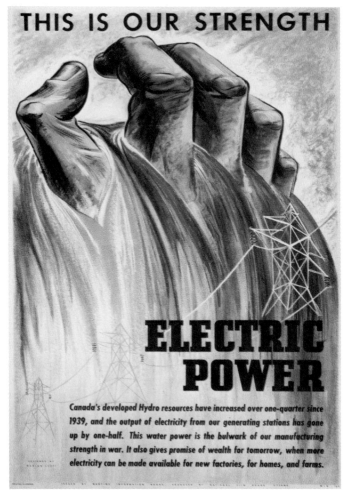

THIS IS OUR STRENGTH

ELECTRIC POWER

Canada's developed Hydro resources have increased over one-quarter since 1939, and the output of electricity from our generating stations has gone up by one-half. This water power is the bulwark of our manufacturing strength in war. It also gives promise of wealth for tomorrow, when more electricity can be made available for new factories, for homes, and farms.

This wartime poster dramatically illustrates the importance of the build-up of Canada's hydro-electric capacity both for the war effort and the subsequent peace. The poster was designed by Marion Scott and produced by the National Film Board. (Public Archives Canada C-87493)

Hydro-electric Power

With the outbreak of war, a fivefold increase in smelting capacity in Quebec was needed to supply the aluminum needs of industry not only in Canada but also in Britain, the United States, and Australia. Aluminum smelting consumed huge amounts of electricity. To supply this electricity, Shipshaw, one of the world's great hydro-electric power complexes, was built at the confluence of the Saguenay and Shipshaw Rivers.

With high annual rainfall in the area, and a vertical drop of 95 metres along the system, the site was well suited for hydro-electric power development. In addition, a deep fjord where the Saguenay empties into the St. Lawrence provided an excellent natural port for unloading ore and exporting refined metal.

Construction of hydro-electric power plants and aluminum smelting facilities had begun on the Saguenay River in the 1920s. Finished in 1925, the Isle Maligne site had given the Saguenay–Lac St-Jean region the world's largest hydro-electric power installation. Construction on the Shipshaw power project, which encompassed facilities at Chute-à-Caron and at Shipshaw, had also been started during this period, and the first 65 000-kilowatt unit at Chute-à-Caron had begun generating power in January 1931. But plans for further expansion had been cut short by the Depression.

With the outbreak of war, a rapid and massive increase in hydro-electric generating capacity on the Saguenay became imperative. Design responsibility for the Shipshaw project was awarded to H.G. Acres and Company, a Niagara Falls consulting firm. The Foundation Company of Canada Limited was chosen as general contractor.

The speed of construction on the Shipshaw site was remarkable. Only 18 months elapsed from the ground breaking in June 1941 to the production of power by the first two generators in November 1942. Before the

end of 1943, all 12 generators were operational, producing a total of 896 000 kilowatts. Moreover, unlike much wartime construction, the 256-metre-long Shipshaw power station was attractively designed and built to last.

Taken alone, Shipshaw was an impressive feat, though it was just one of several hydro-electric projects that would be developed to supply the wartime needs of the aluminum smelting industry. Smelter capacity was also rapidly increased at nearby centres such as Arvida. And 320 kilometres north of Lac St-Jean, a dam was built at Lac Manouane to conserve spring flood waters for power generation. Completed in 1941, the

Built between 1923 and 1925, Isle Maligne on the Saguenay River in Quebec was the world's largest hydro-electric power installation. At the time of its construction, the Saguenay was a remote region. Camps and housing for hundreds of workers and a new railway for supplies had to be built. The installation's 150 000-kilowatt capacity was double the requirement of the newsprint mill it was built to supply. By the time of its completion, however, a contract for the sale of Isle Maligne power had been signed with Alcoa, the Aluminum Company of America. As a result, the Saguenay region became a major centre for aluminum production. (Alcan Aluminium Ltd.)

dam stood 10 metres high and stretched for 610 metres. It added 2.3 million cubic metres of usable water to the province's water power resources, and thereby over 454 000 kilograms of aluminum to the war effort.

The Passes Dangereuses dam, at Chute-des-Passes on the Péribonca River, was another major wartime project built to help supply hydro-electric power to the aluminum industry. Work started in the summer of 1941 and the dam was completed before the spring floods of 1943. The structure was a concrete gravity dam, 48 metres high and almost half a kilometre long. It created a reservoir measuring some 32 000 hectares with a capacity of 5.2 billion cubic metres – larger than the usable storage capacity of Lac St-Jean. Access to the site required building more than 91 kilometres of road, about 20 per cent of it over muskeg, and nine bridges, the longest of which measured 128 metres.

The production chain that led from massive hydro-electric projects in remote areas of Canada through aluminum production and ended with military aircraft is

Harnessing the great hydro-electric potential of the Saguenay River often required ingenious engineering. The problem of closing the massive concrete gravity dam at Chute-à-Caron in 1930 was solved with a huge, 10 000-tonne, concrete obelisk that was used to stem the river's flow. The obelisk was precast to the contours of the river bottom. Tipped by a dynamite blast, it fell exactly into place. (Alcan Aluminium Ltd.)

Plans for the Shipshaw power plant on the Saguenay River, that were shelved during the Depression, were reactivated with the outbreak of war. The power was urgently needed to supply the nearby aluminum smelting town of Arvida. Shipshaw was built in a record-breaking 18 months during 1941 and 1942. (Alcan Aluminium Ltd.)

just one of many examples of Canada's vital role in supplying manufactured goods and strategic materials for the war effort. Other examples show Canadian engineers and scientists involved in tooling up for production in areas of manufacture much newer to this country and of increasing urgency in the conduct of the war – the making of synthetic rubber, plastics, and armour plate steel.

New War Demands, New Responses from Canadian Industry

Twentieth century warfare elevated petroleum and its derivatives to strategic materials. With its emphasis on mechanization and mobility, the Second World War relied heavily on petroleum to supply fuels and lubricants. However, when Japanese advances in Asia cut off sources of natural rubber, petroleum became even more crucial to the Allied war effort.

Synthetic Rubber

Shortly after Pearl Harbor, plans were made to produce a butadiene-styrene copolymer commonly known as synthetic rubber GR-S. Standard Oil Company, the U.S. parent company of Canada's Imperial Oil Limited, held the patent for the process, and it was decided to build 19 plants in the U.S. and one in Canada.

In Canada, Polymer Corporation Limited was established as a crown Corporation on February 13, 1942. The corporation's goal was to produce crude synthetic rubber for tire factories and other rubber fabricators. Construction on a 53-hectare site in Sarnia, Ontario, began in August 1942, and in September 1943 the first shipment of GR-S rubber left the plant.

The Canadian plant had a wider range of capabilities in converting petroleum-based raw materials into synthetic rubber than its U.S. counterparts. As well as butadiene and styrene for GR-S, it produced isobutylene

The Polymer Corporation plant in Sarnia, Ontario, was rushed to completion under wartime conditions in 1942 to provide crude synthetic rubber for tire factories and other rubber fabricators. Although construction of the huge, 53-hectare complex posed many problems, the first shipment of rubber left the plant only 13 months after construction began. The plant played a crucial role in providing wartime materials and in the rapid growth of the plastics industry after the war. (Malak/Public Archives Canada PA-144680, PA-144677)

for butyl rubber or GR-1, an isoprene-isobutylene copolymer.

Both in the speed of its construction and in its size, the Polymer Corporation plant was a monumental achievement. Construction involved 10 engineering firms, four contractors, and over 5000 workers during peak periods. The complex consisted of "eight acres [three hectares] of actual buildings . . . six miles [10 kilometres] of sewers, five miles [eight kilometres] of roads, one hundred and twenty-five miles [200 kilometres] of piping, and the largest powerhouse in Canada."[36]

The site, in what local residents refer to as Chemical Valley, provided ready access to raw materials and transportation. It was close to a major refinery which supplied the petroleum cracking gases needed as raw materials. Well-developed water and rail transportation facilities brought benzene supplies from the coke ovens of steel mills in Hamilton and Sault Ste. Marie and carried the finished rubber to plants in Ontario and Que-

bec. The St. Clair River supplied over 455 000 kilolitres of water a day to meet the plant's prodigious cooling requirements. The salt brine needed by the plant came from local salt wells.

Overall, and despite the fact that construction was rushed through to completion under wartime conditions, Polymer Corporation's mammoth complex at Sarnia was well designed and built. The synthetic rubber it produced contributed significantly to the war effort.

Plastics

Another strategic material which rose to importance during the war was plastic. Plastics served as a substitute for scarce materials such as brass, aluminum, and rubber, and in many instances they reduced the size and weight of products previously made from these materials and shortened production times.

Stimulated by the demands of war, Canadian manufacturers and engineers rapidly developed a plastics industry capable of producing large quantities of plastic products covering a wide range of uses.

During the early 1940s, however, polyvinyl chloride (PVC) was regarded primarily as a rubber substitute. With the shortage of rubber, PVC was in great demand, particularly in the United Kingdom where it was used in electric wiring, waterproof clothing, groundsheets, upholstery, and flexible tubing.

As a result of this urgent need, Shawinigan Chemicals Limited and Union Carbide Corporation formed Canadian Resins and Chemicals Limited. The new company used Carbide technology to produce PVC, polyvinyl acetate, and copolymer-type resin. By 1944, Canadian Resins and Chemicals was shipping up to 182 000 kilograms of plastics and chemicals to the United Kingdom every month.

As the war progressed, plastics assumed ever-increasing importance, particularly in the aircraft industry. Acrylics were used in airplane noses, gun turrets, domes, windows, and escape hatches; phenolic laminated fabrics were used in aircraft control systems for pulleys, cable airleads, and control tabs; plastics or resin-bonded plywood and laminates were used for auxiliary gas tanks, flooring, bulkheads, and seats.

Like airplane manufacturers and ship-builders, the rail car and automobile industries converted their factories to wartime production. In the photograph on the left taken in 1941 or 1942, workers at the Ford Motor Company of Canada Ltd. in Windsor, Ontario, assemble armoured Universal Carriers. The 1942 photograph to the right shows tank assembly at the Angus rail car shops in Montreal. The photograph was used for war propaganda and the white lines enclose the portion printed for this purpose. (NFB/Public Archives Canada PA-132352, PA-116370)

The army and navy also made extensive use of plastics. Among the more demanding challenges from an engineering point of view was the substitution of plastics for metal in the No. 69 hand grenade and the No. 247 fuse. Here, plastics had to be treated as a precision material and moulded to precise specifications.

Armour Plate Steel
Just as the wartime demand for airplanes stimulated aluminum production, the need for large numbers of land and water vehicles sparked rapid growth in the Canadian steel industry.

Prior to 1940, all armour or bulletproof plate used in Canada had been imported from the United Kingdom and the United States. However, to fulfil its wartime obligation to manufacture vehicles such as Universal Carriers, tanks, and armoured cars, Canada now needed to produce its own armour plate and armour castings.

Facilities at Dominion Foundries and Steel Limited in Hamilton were converted to the production of armour steel. The company sent out its first shipment in late 1940 and by the following year was producing up to 91 tonnes a day.

and auxiliary vessels, approximately 3300 special-purpose craft, and thousands of smaller craft.

In addition to consuming large quantities of steel, naval construction required that steel plate be made to exceedingly precise specifications. At the beginning of the war, the Canadian steel industry was incapable of producing steel in the specific thicknesses required. In 1942, the Steel Company of Canada in Hamilton built an entirely new mill to meet this need.

The War's Impact on Canadian Industry

During the war years, corporations involved in strategic industries grew rapidly, and regions of the country which had previously depended primarily on resource extraction acquired an industrial infrastructure. But wartime censorship meant that Canadians remained largely unaware of the fundamental economic changes that were swiftly transforming their country.

However, even when armour plate steel was available, Canada still lacked planers and other machine tools to work the steel into vehicle parts. Skilled labour was also in short supply. But Canadian engineers devised a number of ingenious methods that would reduce dependence on scarce labour and equipment.

On the Valentine tank, for example, a part formerly made of machined armour plate was replaced by a casting. This saved 100 hours of machine time per vehicle – a considerable cost-saving in an overall production run in Canada of more than 1000 tanks.

It was Canada's wartime shipbuilding activities that placed the greatest demands on the steel industry, however. Modern naval ships require vast quantities of steel. For example, the 58 corvettes that Canada had built by the end of 1940 each required 635 tonnes of steel plates and shapes.

But the corvettes represented only a tiny fraction of Canadian shipbuilding production. Before the war, shipbuilding employed about 4000 people; by 1943, it was employing over 126 000 in 25 major and 65 smaller yards. By war's end, Canada had produced 487 escort ships and minesweepers, 391 cargo vessels, 254 tugs

After the war, however, everyone began to take stock. A number of publications appeared recounting the efforts Canada had made to supply vital materials and products for the Allied cause.

Developments in the province of British Columbia illustrate the type of economic impact the war had on virtually every region of the country. In December 1945, the Sun Publishing Company of Vancouver published *Industrial British Columbia: Canada's Magnificent New Industrial Empire Geared to Efficient High-Quality Production Ready for Immediate Post War*

Reconstruction.[37] In the foreword, C.D. Howe, then Minister of Reconstruction, wrote, "The development and expansion of Canadian industry during the past six years is now part of the history of Canada's war effort. In no province were our accomplishments on the industrial front of higher order than in the Province of British Columbia."[38] The publication recounted the contributions that British Columbia industries had made to the war and the roles they hoped to assume in peacetime.

Photographer Grant Gates, a Stelco engineer, was attracted by the aesthetic possibilities of industry. One of his favourite subjects was the Stelco plant in Hamilton during the 1930s and 1940s. The photograph to the left, showing the Stelco blast furnace, he titled "Old Faithful." Above, molten steel is being poured. The photograph on the right that shows the rolling mill, Gates titled "A Ship in the Making" to emphasize the importance of the steel industry to the war effort. (Public Archives Canada PA-138557, PA-147982, PA-7983)

For example, during the war, Canadian Forest Products Limited of New Westminster had manufactured edge-grain spruce and birch plywood for Mosquito bombers and trainer planes. Now it was making hotplate-pressed waterproof and water-resistant Douglas fir plywood for civilian uses.

A similar pattern was repeated in a number of other industries. Because of its long history of shipbuilding, British Columbia often called itself the "Clydeside of Canada." The industry had expanded significantly during the war and now hoped to continue building vessels for peacetime use.

Vivian Engine Works Ltd., which had been manufacturing gasoline engines since 1909 and diesels since 1932, had similar aspirations. The company had set up Vivian Diesels and Munitions Ltd. during the war with Canadian government assistance. The new company had produced 634 diesel engines for wartime use in powerboats, generator sets, and pumping units. It had also made precision instruments for a variety of military applications, including gunsights and fusesetting machines for four-inch anti-aircraft guns.

In addition to creating whole new industries and expanding others in the various regions of Canada, the war effort had a major impact on individual large corporations. This was well illustrated in another 1945 document, *Of Tasks Accomplished*, published by the nationwide Dominion Bridge Company Limited.[39]

Dominion Bridge had manufactured an impressive number of products during the war, including bridges, buildings, cargo ships, cranes, engines, boilers, condensers, and pressure vessels, as well as various types of ordnance – such as shells, cartridge cases, and the sophisticated Vickers Mark VIII two-Pounder anti-aircraft gun.

The Vickers Mark VIII provides an excellent example of the high level of production and management skills that Canadian engineers demonstrated during the war. Dominion Bridge was in fact the only manufacturer of the Vickers Mark VIII outside the United Kingdom. One of the most effective anti-aircraft weapons in the war, it was also an extremely complicated piece of equipment – each gun consisted of approximately 600 parts.

Assembly engineers at Dominion Bridge concentrated on developing a manufacturing process for the Vickers Mark VIII that would minimize the need for scarce, skilled labour and shorten production times. To ensure a balanced flow of parts, they prepared components in lots. Then, as much as possible, they broke the work down into simple, easily learned operations, reserving skilled labour for essential tasks. The engineers also used a variety of fixtures, jigs, and gauges to reduce set-up times and ensure continuous accuracy. Within a year, the system had shortened manufacturing time for many components by 20 per cent.

The wartime activities of Dominion Bridge were carried out in locations across the country. The company had plants and wholly owned subsidiaries in Amherst, Lachine, Ottawa, Toronto, Sault Ste. Marie, Winnipeg, Calgary, and Vancouver.

In Dominion Bridge, as in many other corporations, the massive scale of wartime projects required cooperation among plants all across Canada, which in turn helped foster regional economic interdependence and a growing sense of national identity.

Overall, the war effort linked Canadians as never before and accelerated development in many sectors of the economy. In showing its ability to gear up strategic industries to supply vital materials and manufactured products, Canada demonstrated to the world the high quality of its engineering capabilities and industrial resources.

6

Postwar Boom to Expo 67

During the Second World War, Canada had proven its industrial and engineering capabilities as never before. New industries had been created; new materials and processes developed or improved. A nation that had survived a decade of depression followed by six years of war had reason to be confident of its resilience and strength.

Yet, Canadian optimism was tempered with an underlying fear of the unemployment and recession that had followed wars in the past. War workers who had manufactured products now no longer required, returning veterans, and the numerous refugees and displaced persons from war-torn countries all needed to be integrated into peacetime society.

There was, in fact, a good deal of work for these people to do. The Depression and the war years had slowed or halted the building and repair of roads, water, sewer, energy, and communications facilities. Catch-up construction provided work for hands left idle at the end of the war. The country and its engineers rose to the challenge of modernization with such massive projects as the Trans-Canada Highway, the St. Lawrence Seaway, the Trans-Canada Pipeline, and a trans-Canada microwave system.

Canada's First Coast-to-Coast Highway

As the war wound down, many provincial governments made plans for postwar road construction. In British Columbia, for example, shortly after the war, two major projects were started with a combined cost of approximately $10 million – the Peace River Highway and a road link between Hope and Princeton.

Recognizing the need for a national highway system and for the coordination of provincial road building efforts, the Parliament of Canada on December 10, 1949, passed "an act to encourage and assist in the construction of a Trans-Canada Highway."

The Trans-Canada Highway was a significant achievement on a number of levels. Stretching 7821 kilometres, it was the longest paved highway in the world. It was built to uniform standards of right-of-way, curvature, gradient, sight distance, pavement, shoulders, and bridges. By the end of 1970, the Trans-Canada Highway had cost close to $1.5 billion, with over half of the cost being borne by the federal government.

The highway was also a notable political accomplishment. Construction required the cooperation of the 10 provincial governments and the numerous municipalities along its route. The 1949 act was, in fact, the first federal–provincial agreement under which all of the provinces agreed to uniform conditions.

As an engineering achievement and in terms of its importance to the country, the Trans-Canada Highway rivals the Canadian Pacific Railway. Indeed, engineers on the highway confronted the same challenging Canadian geography as those who had built the transcontinental railway.

The problems of the Precambrian Shield in central Canada with its hard rock, bogs, and numerous lakes, of heavy gumbo clay on the Prairies, and of muskeg in northern Ontario and Newfoundland all had to be overcome. But the high mountains of Alberta and British Columbia with their potential for avalanches and their deep river chasms pose perhaps the greatest challenge for any transportation route across Canada.

Between Golden and Revelstoke, B.C., the Trans-Canada Highway traverses Rogers Pass where snowfall averages 8.6 metres a year and avalanches are common. A research team was sent into the field to identify the hazardous avalanche areas and develop methods of protecting motorists.

The means they devised included conical earth mounds built to act as obstacles to the flow of rock, earth, snow, and ice; dams to change the course of

Between Golden and Revelstoke, British Columbia, the Trans-Canada Highway traverses Rogers Pass in the Selkirk Mountains. Builders of the highway through the pass had to cope with the same heavy snowfalls and avalanche hazards that had hampered construction of the Canadian Pacific Railway in the 1880s. (Douglas Leighton)

Different eras of bridge building in British Columbia are reflected in these two versions of the Alexandria Bridge over the Fraser River. The 1926 suspension bridge in the foreground replaced the original 1863 structure on the Cariboo Road. The newer bridge in the background was built in 1961 as part of the Trans-Canada Highway. Its elegant arched span was designed by A.B. Sanderson, who used computer modelling to test stresses. (Province of British Columbia 14-00278)

Constructed on a cost-sharing basis by the
provinces and the federal government, the
7821-kilometre Trans-Canada Highway
opened in July 1962. This photograph,
taken near the Second Meridian in Saskatchewan in 1952, shows movable rock
crushing equipment of the type that
followed along as the highway progressed.
(Public Archives Canada PA-111584)

avalanches; and bench defence systems which catch and hold snowslides. Permanent emplacements were built for mortars which are fired to trigger avalanches before they become dangerous. In areas where slides are unavoidable, 8230 metres of snow sheds were built. Avalanches run over the snow sheds and continue harmlessly along the other side of the highway.

The Trans-Canada Highway marked the first time these protective measures had been used in North America. They resulted in a highway through the mountains that was significantly safer than the alternative routes.

Deep river chasms in the mountains posed another major obstacle for the highway. A number of fine bridges were built, such as the Port Mann Bridge over the Fraser River near New Westminster, which represented the first use of orthotropic steel decking in North America. It is a beautiful bridge formed by a stiffened tied arch of three spans of 107, 366, and 109 metres which made it, when it opened, the longest of its type in the world.

The Louis-Hippolyte Lafontaine bridge-tunnel near the east end of Montreal is another highlight of the Trans-Canada Highway. Motorists leaving Montreal Island go under the main channel of the St. Lawrence River by tunnel, emerge on Charron Island, and then

Completed in the summer of 1960 as part of the Trans-Canada Highway, the Hugh John Fleming Bridge over the St. John River, New Brunswick, was a milestone in concrete arch bridge design and construction. It made innovative use of wide-span parabolic arches that gradually increased in height over a one per cent grade. An alternative proposal for a less costly steel-deck truss bridge was rejected in favour of this outstanding concrete design. (Canada Cement Lafarge Ltd.)

The St. Lawrence Seaway has received worldwide recognition as one of the most important man-made waterways. More than $1 billion was invested and over 500 U.S. and Canadian engineers were involved in the massive project. These three pictures show the St. Lambert Lock. Located at the southern end of Victoria Bridge in Montreal, St. Lambert is the first lock on the St. Lawrence Seaway. The lock, which opened in April 1959, lifts ships 4.6 metres from Montreal Harbour to the Laprairie Basin. To provide clearance for ships, the trestle in the photograph below and a similar one at the other end of the lock were replaced by lift spans. (St Lawrence Seaway Authority 7417, 5499; National Film Board 66-14044)

proceed by bridge across the remaining width of the river to the south shore at Longueuil.

The Louis-Hippolyte Lafontaine bridge-tunnel traverses a total water distance of 1966 metres. The largest part of this is the 1390-metre tunnel which houses a six-lane highway under the main channel of the St. Lawrence. The tunnel is composed of seven huge identical sections, each of which is 91.4 metres long, 36.6 metres wide, and 7.9 metres high.

The tunnel required innovative engineering on an unprecedented scale. The sections were made of rein-forced concrete which was poured on land while a trench was being cut in the riverbed. When the trench was ready, the sections were floated into the river, sunk into position in the trench, and then joined together to form the tunnel.

The tunnel is an imaginative, daring, and carefully executed piece of work. Moreover, it was an important engineering landmark. Aside from its technical merit, it brought important professional recognition to Canadian engineers such as Roger Nicolet, Armand Couture, and Bernard Lamarre. The latter was then in the process of

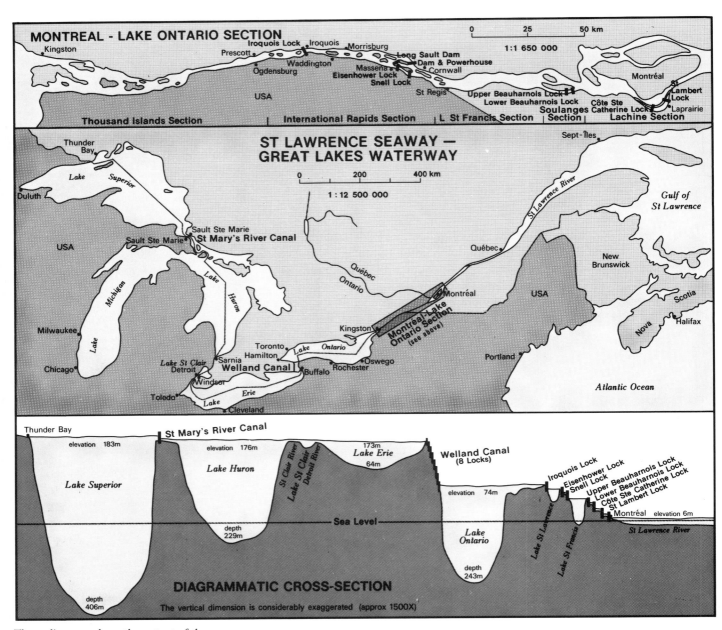

MONTREAL - LAKE ONTARIO SECTION

1:1 650 000

Kingston • Iroquois Lock • Iroquois • Morrisburg — Long Sault Dam — Dam & Powerhouse — Cornwall
Prescott • Waddington • Massena • Eisenhower Lock • Snell Lock • St Regis
Ogdensburg
USA
Upper Beauharnois Lock • Lower Beauharnois Lock • Côte Ste Catherine Lock • Soulanges Section • Montréal • St Lambert Lock • Laprairie

Thousand Islands Section | International Rapids Section | L St Francis Section | Lachine Section

**ST LAWRENCE SEAWAY —
GREAT LAKES WATERWAY**

1:12 500 000

Thunder Bay
Lake Superior
Duluth
Sault Ste Marie • St Mary's River Canal
USA
Lake Michigan
Lake Huron
Milwaukee
Toronto • Hamilton • Lake Ontario
Chicago
Lake St Clair • Detroit • Sarnia • Welland Canal • Buffalo • Rochester • Oswego
Windsor
Toledo • Lake Erie
Cleveland
Kingston • Montréal-Lake Ontario Section (see above)
Québec Ontario
Québec
St Lawrence River
Sept-Îles
Gulf of St Lawrence
New Brunswick
USA
Nova Scotia • Halifax
Portland
Atlantic Ocean

DIAGRAMMATIC CROSS-SECTION
The vertical dimension is considerably exaggerated (approx 1500X)

Thunder Bay
elevation 183m
Lake Superior
depth 406m

St Mary's River Canal
elevation 176m
Lake Huron
depth 229m

St Clair River • Lake St Clair • Detroit River
Lake Erie 173m 64m

Welland Canal (8 Locks)
elevation 74m

Lake Ontario
depth 243m

Sea Level

Iroquois Lock • Eisenhower Lock • Snell Lock • Upper Beauharnois Lock • Lower Beauharnois Lock • Côte Ste Catherine Lock • St Lambert Lock
Lake St Lawrence • Lake St Francis
Montréal elevation 6m
St Lawrence River

These diagrams show the extent of the
St. Lawrence Seaway and the dramatic
changes in elevation from Thunder Bay on
the northern shore of Lake Superior to Sept-
Îles at the mouth of the St. Lawrence River.
(*The Canadian Encyclopedia*)

propelling one of Quebec's many engineering firms into one of the world's largest and most respected.

Among the many other engineering highlights of the Trans-Canada Highway is the 1219-metre Canso Causeway which links Cape Breton Island with the Nova Scotia mainland. Completed in 1955, the causeway carries both rail and highway traffic and has freed ports along the Canso Strait to year-round, ice-free Atlantic shipping.

The Trans-Canada Highway, like the railroads in the nineteenth century, with their brilliant and innovative engineering accomplishments, opened up previously inaccessible parts of the country. It provided a reliable trucking route which lowered the costs of many goods, and made other goods, such as fresh fruits and vegetables in winter, available for the first time. Primarily, it linked together the people of a vast, sometimes forbidding, and sparsely populated land.

A Gas Pipeline Links West to East

The Trans-Canada Pipeline owes its existence to C.D. Howe, federal Minister of Trade and Commerce in the Liberal administration of Prime Minister Louis St. Laurent. Howe felt that the huge reserves of natural gas in western Canada and the correspondingly huge markets for gas to be used in heating and as a chemical feedstock for industries in central Canada made a pipeline linking the two regions a national necessity.

In 1954, Howe put together a syndicate of private American and Canadian businessmen. The syndicate had the specialized products, the expertise, and much of the capital to build an all-Canadian pipeline.

On May 8, 1956, Howe introduced a bill in Parliament which would authorize the construction of the pipeline and guarantee federal loans for part of its costs. If construction was not to be delayed a year, it would

have to start in early June. This meant that the bill required quick passage.

The contentious financial arrangements and the possibility of permanent United States domination of the pipeline company led to one of the most famous and vitriolic debates in Canadian parliamentary history. On June 6, the Liberal government cut off debate by invoking closure – at that time, a controversial and rarely used measure. The debate and the way in which the government had ended it were important factors in the defeat of the Liberals in the general election the following year.

Nevertheless, by October 1958, the 3700-kilometre pipeline between Burstall, Saskatchewan, and Montreal was completed. Despite the controversy surrounding its creation, the pipeline was a major engineering feat. Its construction required that engineers surmount many of the same obstacles faced in building the railroads and the Trans-Canada Highway – cold weather conditions, the rugged Precambrian Shield, vast stretches of muskeg, and dense forests.

Nation-wide Microwave Transmission Revolutionizes Communication

In addition to the need for improvements in transportation and energy, Canada's tremendous postwar economic growth created demands for updating and expanding the country's telecommunications systems.

As telephone traffic volumes grew, it became increasingly apparent that the standard line system would not be able to keep pace. Moreover, the Canadian Broadcasting Corporation wanted to link the country with coast-to-coast network television programming. Fortunately, the Chairman of the CBC was an electrical engineer, Alphonse Ouimet, who fully understood both the workings and the potential of microwave transmission. Under Ouimet, it was agreed that a nation-wide,

standardized microwave transmission system would be the most viable solution to meet both these requirements.

During the war, the first microwave corridor was surveyed between Montreal and Windsor. Shortly after the war, microwave communications systems had begun to link various parts of the country. In 1948, the world's first microwave connection for both commercial and voice transmission replaced an unreliable underwater cable between Prince Edward Island and Nova Scotia. In 1953, Bell Canada established a microwave system for telephone communications between Montreal, Ottawa, Toronto, and a year later, Quebec. It also extended the chain from Toronto to Buffalo for television.

Early in 1954, the CBC called for tenders on a national microwave system. The contract was awarded to the TransCanada Telephone System, a consortium of seven telephone companies: Bell Canada, the Maritime Telegraph and Telephone Company, the New Brunswick Telephone Company, the Manitoba Telephone Company, Saskatchewan Government Telephones, Alberta Government Telephones, and the British Columbia Telephone Company.

The Trans-Canada Microwave Radio Relay System was to be the longest microwave system in the world. The immensity of this engineering challenge was complicated by the fact that microwave transmission requires a direct line of sight between repeaters which must be spaced an average of 50 kilometres apart.

Although much of Canada had been mapped, the mapping had of course not been done with a series of coast-to-coast sight lines in mind. In addition to carrying out aerial mapping, crews were sent for on-the-spot inspections of potential sites. Sites were examined and assessed for the incidence of obstructions such as trees, high land, or buildings that could interfere with trans-

mission; access roads; and the availability of power and the costs of bringing it to the site.

Parabolic test antennae atop portable aluminum towers which could be erected to a height of 61 metres within 24 hours were used to conduct transmission tests at potential sites. One surprise was the particular transmission problem encountered on the flat prairies. The countless sloughs acted as reflectors which sent the microwaves off in unexpected directions. Careful planning was required to avoid "ghost" signals reflected from the sloughs.

Instead of transporting heavy test equipment up to potential sites in the mountains in British Columbia, engineers used a simple but ingenious method to establish direct lines of sight. After ascertaining their locations with walkie-talkies, they then used ordinary hand mirrors to reflect the sun across to the neighbouring crew. In this way, a direct line of sight free of obstructions between microwave tower sites was established.

Once the sites had been selected and purchased, the next step was the often arduous task of building access routes for the heavy equipment needed for construction. At Cresston, British Columbia, the access road had a prohibitively steep 20 per cent grade and 33 switchbacks in 3.2 kilometres. In the northern Ontario section between Long Lac and Hearst, many of the roads to tower sites had to be bulldozed through the bush.

In other areas, building a road was simply out of the question. Near Hope, British Columbia, engineers built

The snow and ice on this microwave transmission tower on Dog Mountain in British Columbia illustrate the extreme weather conditions which made building and operating these towers difficult. (B.C. Tel)

a 3597-metre-long aerial tramway to lift workers and materials through 1341 metres at speeds of between eight and 16 kilometres an hour.

The construction of the microwave facilities themselves presented major engineering challenges. In addition to foundation work and small buildings to house radio and associated equipment, each site required a steel tower ranging from 12 to 204 metres and weighing as much as 109 tonnes.

On the prairies, the gumbo soils, famous for their capacity to bog down wheeled vehicles, presented a serious problem. These soils remain liquid for a metre or more below the surface. To provide a stable foundation, footings were placed well below the frost line. Each footing was bound to four, 4.6-metre-long, reinforced concrete piles erected in holes in the gumbo.

On July 1, 1958, a 90-minute CBC national television special, co-hosted by broadcaster and future premier of Quebec René Lévesque, celebrated the successful completion of the trans-Canada microwave system. The world's longest, the new system represented a major achievement for Canadian engineering. Much of the equipment had been developed specifically for the project, with engineers in the field only a few steps ahead of those in the laboratories, design offices, and manufacturing facilities.

The microwave system was also an important step in the continuing development of Canada's communication facilities. It provided a nation-wide, east-west communications channel capable of transmitting high volumes of telephone and television traffic from Charlottetown to Victoria and all major cities in between.

The microwave network formed the nation's communications backbone. Even before the system was fully operational, tropospheric scatter systems were put in place to augment it by extending telephone service

60 Canada
Postes
Postage

CF-SAM

Noorduyn Norseman

Prospecting and mining in the Canadian north during the 1930s created a demand for a plane able to reach wilderness regions. The Norseman met this need. Its importance in Canadian history was recognized by Canada Post with a stamp from a painting by Bob Bradford. The first Norseman was built in Montreal in 1935 by a small team headed by Robert Noorduyn, a Dutch immigrant who had designed a Fokker freight plane in Europe. The Norseman was a versatile and rugged bush plane, able to land on wheels, floats, or skis. It had a high wing for stable flight and easy parking near docks; a large tail for good control on water; and huge removable doors for bulky freight. It was also a comfortable plane, with upholstered chairs and aluminum foil insulation for warmth and sound reduction. The photograph below shows Norseman assembly at the Noorduyn Aviation Ltd. factory in Montreal in 1941. During the Second World War, 900 of these planes were built for the allied forces. (Canada Post Corporation; National Aviation Museum 5945)

into remote areas. Because relay stations can be placed as much as 300 kilometres apart rather than the average of 50 kilometres required for microwave stations, tropospheric scatter systems were a more economical solution for supplying communications to sparsely populated areas. Since that time, satellite communications and fibre optics have continued to improve services.

The Rise and Fall of the Avro Arrow

Even before the Second World War, the Canadian aviation industry had begun pioneering the production of planes, such as the sturdy Norseman, designed to meet Canadian needs. The tradition continued after the war with the de Havilland Beaver, a superb workhorse for opening up, servicing, and maintaining contact with northern and remote areas inaccessible by road.

By 1945, the 10-year-old Norseman design was showing its age. Following an extensive market survey, de Havilland of Canada designed the Beaver. The first prototype flew on August 16, 1947, and the plane was an immediate success. The Beaver's features included all-metal construction for durability; short take-off and landing distances; simple methods of switching among floats, skis, and wheels; and doors designed for bulky loads. In the late 1940s, the Beaver won a U.S. Army contract for light transport aircraft. This unusual purchase of foreign

aircraft required a special act of the U.S. Congress. The Beaver has been sold all over the world and remains popular today. (National Aviation Museum 11481, 11487, 13379)

However, during the war, Canada had created a massive industry specializing in the production of high-performance military aircraft. To capitalize on this investment, the federal Liberal government decided to fund the postwar development and manufacture of high-speed, jet-propelled, military aircraft. The industry, it was felt, had a good potential for growth. It would place Canada among the technologically advanced nations of the world and would have numerous scientific, engineering, and economic spinoffs.

The Avro Arrow jet fighter, developed by A.V. Roe of Canada, was the culmination of this program. It was a technically advanced, supersonic, twin-engined, all-weather, long-range interceptor.

However, in 1959, the Conservative government, which had taken over from the defeated Liberals in 1957, cancelled the Avro Arrow program and stopped all work on it. This decision, which threw approximately 14 000 skilled aircraft personnel out of work, was highly controversial and is still vigorously debated.

The government argued that the program had been unwise from the beginning. Given Canada's relative population size, small military budget, and the fact that other governments tended to favour their own domestic suppliers for military hardware, the Conservatives believed that sales could never justify the investment. It was time, they felt, to minimize the losses and get out.

Whether the government's decision was wise or not, it did wipe out over a decade's work and almost an entire industry. Moreover, the government not only terminated the program, it also systematically destroyed the records of the project's engineering research and development work. Flight reports, technical data, drawings, and even photographs were destroyed. The 37 planes – both operational and those in assembly – were cut up and sold for scrap.

The decision to destroy the Arrow data and the planes themselves was apparently made to prevent any future government from reversing the decision and reviving the program. Unfortunately, it also wiped out a

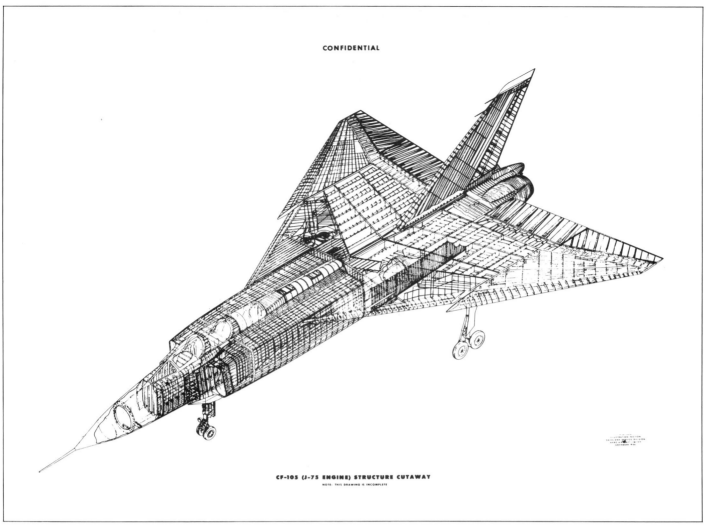

CF-105 (J-75 ENGINE) STRUCTURE CUTAWAY
NOTE: THIS DRAWING IS INCOMPLETE

The Avro Arrow was officially unveiled on October 4, 1957 – the same day that the U.S.S.R. launched its first satellite. The Arrow was the first fighter plane designed for sustained operation at supersonic speeds and made Canada a world leader in supersonic fighter plane design and construction. The Arrow program was cancelled on February 20, 1959, a date that became known as Black Friday in aviation circles. (National Aviation Museum 6891, 6889)

wealth of knowledge which would have been of benefit to others in the aircraft and related industries. Thus, while the government may have been wise to stop costly development work, its decision to destroy the records can only be viewed as shortsighted and ill-advised.

The CANDU Reactor

In contrast with the Arrow, the CANDU (*Can*ada *D*euterium *U*ranium) nuclear reactor is a case where Canada carried through with the development of new technologies created during the war. In 1945, the ZEEP Nuclear Reactor at Chalk River became the world's first nuclear reactor in operation outside the United States. This was followed in 1947 by the NRX, the world's most powerful research reactor, and by the NRU reactor at Chalk River in 1957. These reactors provided the base for the development of fundamental nuclear power technology. Canada's excellence in this field was recognized in the mid 1950s when the NRX and then the NRU were chosen to produce nuclear fuel for the U.S. Navy.

During this same period, Ontario Hydro was looking for new sources of electricity to satisfy the rapidly growing demand in the province. The joint industry–government approach that had proved so successful in Canada in the past was followed. Atomic Energy of Canada Limited (AECL), Canadian utilities, and private industry concluded that the CANDU reactor was the route to pursue.

Perhaps the greatest challenge posed by CANDU was the necessity of adhering to extremely high quality standards in every aspect of the project – in design, manufacture of components, construction, and maintenance and operation. The result was not only successful generating installations of varying sizes up to Ontario Hydro's Darlington Station, the world's

largest, but also advances and activity in numerous other related fields. Perhaps best known are the medical applications, such as the cobalt bomb for radiotherapy. Other spinoffs from the nuclear program include automatic computerized control systems, simulator models, remote handling techniques, and fundamental advances in areas such as metallurgy and chemistry.

The Alouette and Anik in Space

On September 29, 1962, the U.S. National Aeronautics and Space Administration (NASA) launched Canada's first satellite, the Alouette I, and Canada became the third nation in space. The Alouette I was the first satellite designed and built outside the United States and the Soviet Union.

With a weight of 146 kilograms, a height of 0.9 metres and a diameter of 0.9 metres, the Alouette I was relatively large for the period. It was also one of a series of Canadian-built satellites which were unusually complex and sophisticated. Others in the series included Alouette II launched in 1965, ISIS (International Satellites for Ionospheric Studies) I in 1969, and ISIS II in 1971.

The launching of Alouette I in 1962 marked Canada's entry into the space age. Here, the Alouette I is being assembled at the Defence Research Board. It was designed for scientific experimentation and, in this, was typical of the first phase of the Canadian space program. During the 1960s, the emphasis shifted to satellites for surveying natural resources and for telecommunications. (Department of Communications 62-6166)

All of these early satellites were aimed at understanding the upper atmosphere and particularly the ionosphere – the layer of charged particles that reflects shortwave signals and is therefore extremely important in high-frequency radio communications and radar.

The combined Alouette/ISIS program was an immense scientific success – it resulted in the publication of over 1200 papers and reports. It also embodied a number of important advances in engineering.

The satellites were designed and built by the Defence Research Telecommunications Establishment (DRTE) of the Defence Research Board of Canada. Engineering staff at DRTE had to design within severe constraints imposed on size, weight, and power consumption. To send and receive signals, the satellites needed to deploy large antennae that would function reliably in a distant and hostile environment.

To meet the design goals and constraints, innovative components were developed including high-reliability solid-state electronics, multi-octave HF antennae, high-power wide-band transmitters, and a long-life rechargeable power supply.

The power supply consisted of 6500 solar cells delivering 23 watts at the start of their life and six 15.6-volt, 5-amp-hour nickel cadmium batteries which lasted for 10 years. The nickel cadmium batteries were the result of a major development effort undertaken by the Defence Research Board's Chemical, Biological and Radiation Laboratories aimed at improving the reliability of commercial batteries. NASA studies found the Canadian nickel cadmium batteries superior to any others available at the time.

The principal experiment on the satellites was swept-frequency radar covering the 1–12 megahertz band. Most satellite instrumentation available at the time could probe only its immediate environment. The radar used for sounding on the Alouette/ISIS program could explore the entire ionosphere from about 300 kilometres up to the satellite's altitude of about 1000 kilometres.

The radar system relied on extendible, dipole antennae – the most important and innovative aspect of the satellites. On the ISIS I and II satellites, the extended antennae measured 73 metres tip-to-tip.

The antennae consisted of four separate arms each known as a Storable Tubular Extendible Member (STEM). The production of a device of such length which would fit inside a satellite approximately one metre across, withstand the stresses of the launching operation, and then extend and operate in the forbidding environment of space represented a fundamental step forward in the engineering of complex movements.

The basic mechanism developed for this purpose has continued to evolve and was used, most notably, in the Canadarm developed by Spar Aerospace Limited for the NASA Space Shuttle.

Both the Alouette/ISIS satellites and the Canadarm demonstrated what has become a mark of Canadian aerospace engineering – reliability. The Alouette I, for example, set a record for operational longevity.

The reason for this remarkable reliability is the approach taken by Canadian engineers. Rather than relying on precise tolerances and parts with low failure rates, they have tended to engineer for wide operating margins. Systems are designed to continue functioning under extreme conditions and even when individual components malfunction or fail.

In the Alouette satellites, for example, the electronics were designed to tolerate larger-than-forecast variations in radiation, temperature, and power supply, and to cope with a wide range of malfunction and failure. The approach proved its mettle only two months after the launching of Alouette I. Radiation in the upper atmosphere from a United States hydrogen bomb test damaged U.S. satellites while the Alouette I survived unscathed.

The Alouette/ISIS program was aimed at researching properties of the atmosphere important in communications. The Canadian satellite program has continued to be driven by the communications needs of a vast country with a widely dispersed population. In 1964, Canada joined Intelsat, the international satellite system for the exchange of telecommunications traffic. In 1969, Parliament established Telesat Canada Corporation, with a mandate to set up a domestic satellite communications system. This was followed in 1972 by the launching of the first Anik satellite, making Canada the first country in the world with a domestic communications satellite operating in geostationary orbit.

The performance of the Alouette/ISIS satellites and their successors gave Canada an international reputation for excellence in satellite design and engineering, and led to the creation of an active aerospace industry. Over the past 25 years, the industry has grown at a rate of more than 20 per cent a year and it now employs over 3500 people. In 1985, Canada's aerospace industry sales exceeded $350 million, of which exports accounted for more than 70 per cent.

Engineering an End to the Winnipeg Floods

The Red River which meanders through Winnipeg is one of Canada's great rivers. Slow-moving and silt-laden, it provided for centuries an important transportation route for native peoples, European fur traders, explorers, and settlers, who plied its waters in craft from canoes to steamboats.

The huge drainage area for the Red River covers approximately 12.5 million hectares. Major floods have occurred as long as written records have been kept. The worst of these were in 1826, 1852, 1861, and 1950 when flood waters rose more than nine metres above normal.

In the nineteenth century, Sir Sandford Fleming had advised that the city of Winnipeg be located elsewhere to avoid the flood problem. But Fleming's advice was ignored and, by 1950, when the worst flood in almost a century struck, a major city had arisen on the site.

The Red River flood of May 1950 was one of the costliest disasters in Canadian history. One eighth of the greater Winnipeg area and approximately 145 000 hectares of land were inundated, 10 500 homes were flooded, and 100 000 people were evacuated. Civilian and military forces combined to bring in huge quantities of relief equipment and supplies. For example, over 455 tonnes of freight were airlifted into the area in one 96-hour period ending on May 12. Flood waters finally began receding on May 13 and, when it was over, the damage totalled well over $100 million.

In the wake of the flood, the Government of Manitoba appointed a royal commission to study the problem and propose solutions. In its 1956 report, the Royal Commission on Flood Cost Benefit recommended the construction of a flood control system. The costs of future floods, it said, would be much higher than those of building a control system.

Construction of the Red River Floodway began on October 6, 1962. It was one of the biggest excavation projects in Canadian history and required moving about as much fill as did the Panama Canal.

Approximately 76 million cubic metres of earth were excavated from the 47.3-kilometre floodway channel. The width of the channel at the base ranged from 116 to 165 metres and at the top from 213 to 305 metres. In addition, 43.5 kilometres of dykes were built using 2.3 million cubic metres of fill.

Building the floodway around the city also involved dealing with a wide range of transportation facilities and

The Husky 8 vehicle has been in production by Canadian Foremost Ltd. since 1967. However, the vehicle depicted here is a specialized unit used by the Ministry of Oil in the Soviet Union for the control of oil well fires in remote areas of Siberia. The vehicle is equipped with a Wormald Fire Fighting System that was specially designed for this application. Fifty units were delivered to the Soviet Union in 1986. (Canadian Foremost Ltd.)

public utilities. Crossings and bridges had to be made for railways and roads, and for water, oil, gas, power, and telephone lines.

The Red River Floodway is designed to divide water between the floodway channel and the Red River. When the river reaches a flow rate of 850 cubic metres per second, the inlet control structure begins to divert some of the water to the floodway channel. The flow is controlled by two massive 34.3-metre-long and 10.6-metre-high floodway gates. At its maximum, the system allows 1700 cubic metres per second to flow down the floodway channel and 2000 cubic metres per second to flow through a control structure and then down the river's natural channel. The control structure is flanked by dykes for 9.7 kilometres on one side and 33.8 kilometres on the other.

The inlet control structure was itself a major construction undertaking. It consumed 60 800 cubic metres of concrete and 500 tonnes of reinforcing steel.

Completed in 1969, the Red River Floodway was one of the world's major flood control systems. It represents a continuing Canadian tradition of partnership between government and engineers in major public works built to meet Canadian needs.

Canadian Engineers in World Markets

While major projects such as the Trans-Canada Highway and the space program were occupying Canadian engineers in the postwar period, many important engineering companies were also springing up or continuing to grow, companies that have now matured and diversified. Canadian Foremost Limited, The SNC Group, and Lavalin Incorporated are examples of the many firms that have become international leaders in engineering.

Canadian Foremost Limited

With the expansion of oil exploration and production into northern Alberta and the Northwest Territories, it quickly became apparent that there was an acute need for a vehicle capable of carrying heavy equipment over muskeg. In 1952, Bruce Nodwell of Calgary formed Bruce Nodwell Limited with the purpose of developing such a machine.

In 1956, he succeeded in building a vehicle for Imperial Oil Limited that would move a 4.5-tonne seismic rig over muskeg. Earlier vehicles had had some power train and reliability problems. Nodwell's innovative design allowed him to build the longer and wider tracks needed for more powerful vehicles, and

Canadian Foremost Ltd. developed the Magnum 4 Transporter in 1977 for large load transport in adverse terrain. The vehicle was used extensively during construction of the Alaska Highway Gas Pipeline prebuilt section through the Flathead area of British Columbia. The Magnum 4 is equipped with under-the-deck, articulated steering, giving it a remarkably long, 15-metre deck. (Canadian Foremost Ltd.)

to develop a reliable means for transmitting power to the drive and track systems.

Shortly after his first success, Nodwell built a second vehicle for Shell Canada Limited with twice the carrying capacity. By the 1960s, his machines were widely used in northern oil exploration both in deep snow at low temperatures and over muskeg at temperatures above freezing.

In 1965, Bruce Nodwell and his engineer son, Jack, formed Canadian Foremost Limited. Today, the company is a world leader in the design and manufacture of all-terrain vehicles, and about 80 per cent of its sales are made to foreign customers. The company has diversified out of the oil business and now supplies vehicles for a wide range of customers such as high

performance, all-terrain firefighting equipment to the Soviet Union.

The Canadian Foremost vehicles demonstrate one of the important and recurring areas of Canadian engineering strength – the production of specialized products initially based on Canadian needs and then adapted to take advantage of a larger market. With Canadian Foremost and many other Canadian companies, careful engineering research and attention to reliability have created high-quality basic designs that can be modified to meet a variety of customer needs.

The SNC Group
The SNC Group is one of Canada's most successful diversified engineering firms. The initials stand for

The Manic 5 (now Daniel Johnson) Dam was an outstanding achievement. Completed in 1971 as part of the Manicouagan-Outardes River hydro-electric project, it is the world's largest arch and buttress dam. The construction drawings on this page and the following two pages were done by Lili Réthi. (The SNC Group)

Manic 5
Airview
Looking East

Réthi
May 19th
1965

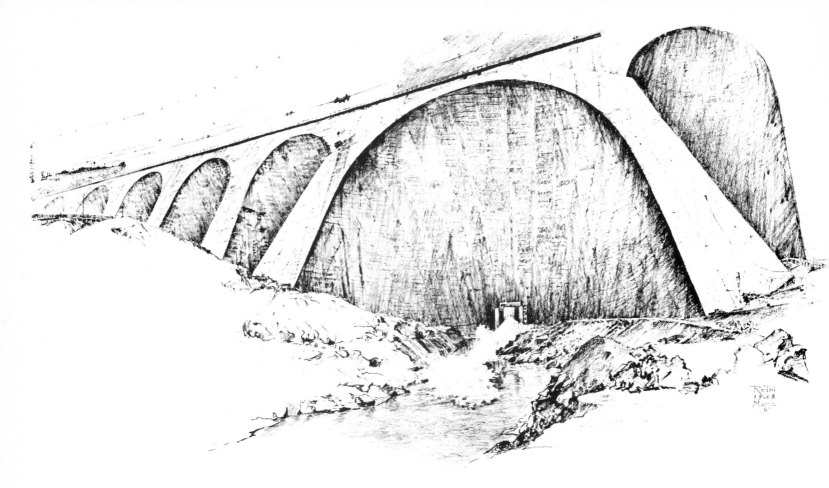

Dr. Arthur Surveyer, Emil Nenniger, and J. Georges Chênevert. Throughout its history, the company has offered a wide range of services. When it began in Montreal in 1911 as Arthur Surveyer & Co., it offered civil, mechanical, and electrical engineering and municipal design and field supervision services. The supervision services have grown into project management for which SNC is famous today.

In its early years, the company gained experience in heavy industry and hydro-electric projects. In the late 1940s, it moved into industrial projects for the textile, food, paper, building material, and chemical industries.

In the early 1950s, SNC was involved in two large metallurgical and mineral projects in Quebec: the Quebec Iron & Titanium's smelter in Sorel, and the Canadian Johns Manville's mill in Asbestos.

During the 1950s, demand for electric power was also increasing rapidly and SNC undertook hydro-electric studies of Quebec's northern rivers. These studies led to SNC's contract with Hydro-Québec for the design of Manic 5 on the Manicouagan River, the world's highest multiple-arch dam and one of the most exciting engineering projects of twentieth century Canada.

Through Hydro-Québec, SNC became involved in the most sophisticated hydro-electric projects in the world. With the expertise and the reputation gained from these projects, SNC and other firms such as Lavalin Inc. were hired for similar projects in other countries. As a result, these firms have continued to grow, prosper, and diversify into the 1980s.

Lavalin Inc.
Among Canadian engineering firms, Lavalin Inc. is preeminent for its work at home and abroad in countries such as the People's Republic of China and the U.S.S.R.

The massive Manic 5 Dam is a daringly engineered concrete structure. It is 214 metres high, 1314 metres long and weighs approximately 6 million tonnes. The dam's relatively thin concrete arch – only 27 metres thick at its base and five metres thick at its crest – is strong enough to hold back a wall of water nearly 183 metres high. (The SNC Group)

Lavalin started in 1936 as Lalonde & Valois, a civil engineering partnership formed by Jean Paul Lalonde and Romeo Valois. At that time, few civil engineering firms were able to deal with soil analysis and geological formation study – two areas of expertise essential to design foundations for increasingly heavy, engineered structures. One year later, NBS was founded to fill this market need and later joined Lalonde & Valois.

Several decades of steady expansion in public works, civil, structural, and foundation engineering followed. This expansion was confined primarily to Quebec and eastern Canada.

The mid 1950s – a very active period for many Canadian engineering firms – marked the beginning of a new phase in the company's development that lasted for two decades. The company grew rapidly as it entered new

engineering fields, such as municipal and sanitary engineering; energy generation and transmission; and mechanical and electrical installations.

Expanding areas of engineering expertise led to work in other parts of Canada and in foreign countries, particularly in west Africa, during this period. The work abroad was important in establishing a base for the company's future expansion in international engineering. In the 1970s, its name was changed to Lavalin Inc., and the former strategy of internal growth was supplemented by a vigorous program of corporate acquisitions.

In combination, internal growth and external acquisitions have produced a staggering list of divisions, areas of professional service, clients, and countries of activity. With recent acquisitions such as those of the Urban Transportation Development Corporation and Cantech, Lavalin has also become a major manufacturer. Manufacturing makes engineering firms less dependent on world construction cycles and provides them with new sources of ideas and experience.

As engineering firms become increasingly larger, more diversified, and visible, their social and cultural policies may be subject to greater exposure and examination. In this regard, Lavalin Inc. – through Bernard Lamarre – has established a national reputation as a patron of the arts.

Expo 67 inspired many original approaches to pavilion design and construction. Montreal architect Moshe Safdie's Habitat (left) took full advantage of construction with prefabricated building components. Canada's Katimavik pavilion, which can be seen in the background below, required four huge spines angled from the base. Made by Dominion Bridge Company, the spines were left to rust and then lacquered. (Museum of Contemporary Photography 66-13286, 67-10471)

Throughout the postwar years, as projects and industries have grown in size and complexity, engineers have assumed an increasingly important role in management. The expertise in planning and managing large projects demonstrated by SNC, Lavalin, and numerous other firms has, in fact, become synonymous with Canadian engineering practice.

The success of Expo 67, which opened in Montreal to celebrate a hundred years of Canadian nationhood, was a tribute to these extraordinary project management skills. The construction of the site was a complex job done in what many regarded as an impossibly short period of time. Managing the project required tight and detailed scheduling so that work could proceed on many fronts rapidly and simultaneously. Without excellent engineering management, Expo 67 could not have been completed on schedule, with resultant dire effects on the outcome of the entire enterprise.

As it turned out, the exposition proved to be a brilliant showcase not only for Canadian scientific and cultural products and their makers, but also for the people whose mind, heart, and vision put it all together successfully – many of them the engineers portrayed in this book.

7
Approaching the Second Century

The passage of time allows historians to place events in perspective, though it is often difficult to evaluate events in the recent past. Yet, it does seem possible to identify some of the events and trends that will figure prominently in future histories of the 1970s and 1980s, and to explore their relationship with engineering achievements.

Major trends include the rapid escalation – and decline – in oil prices and revenues, the movement toward greater equality of the two dominant Canadian language groups, progress toward a fairer measure of social, economic, and political justice for women, and the increasing concentration of population in large urban centres.

Important achievements of Canadian engineering in the 1970s and 1980s include the spectacular CN Tower and the artificial ice islands in the Arctic. Some of them, such as the striking innovations in engineering for the physically disabled, are virtually unknown outside a small circle of specialists.

New Directions in Petroleum Exploration and Extraction

Petroleum is one of the foundations of a modern industrial economy. Crude petroleum supplies the raw materials for products ranging from fuels for engines and energy production to chemical feedstocks, plastics, synthetic rubbers, lubricants, and pharmaceuticals.

During the 1950s and 1960s, oil was cheap and supplies seemed secure. Between 1947 and 1971, for example, the Canadian Consumer Price Index doubled from 100 to 202, while the price of crude oil increased only marginally from $16.00 to $17.22 a cubic metre.

However, the formation of the Organization of Petroleum Exporting Countries and the dramatic rise in oil prices during the 1970s had global consequences. In Canada, the Consumer Price Index increased 2.4 times

between 1971 and 1981, and oil prices shot up nearly sevenfold, to $117.70 a cubic metre.

With the abrupt transition from cheap to expensive petroleum, energy efficiency became a prime concern and, in many instances, an operational engineering goal. In buildings, heating and lighting requirements were reduced. The shift to smaller, fuel-efficient, imported cars threatened the North American automobile industry.

In Canada and other oil-producing areas, the dramatic rise in the cost of imported crude stimulated feverish activity throughout the petroleum industry. Engineers and geologists took measures to increase recovery rates from existing wells. The search for new reserves sent drilling crews into frontier areas in the North and off shore. Renewed efforts were made to extract oil from unconventional sources such as the tar sands.

Oil Sands

From the early days of the fur trade and northern exploration, the Athabasca tar sands of north-central Alberta have fascinated Canadians. As early as the 1790s, the explorer Alexander Mackenzie noted that the naturally occurring pitch could be used for canoe repairs.

The Athabasca tar sands consist of a huge deposit of thick, tar-like bitumen or crude petroleum mixed with sand and other minerals. The bitumen is more viscous at normal temperatures than conventional crude oil and higher in both carbon and sulphur content. Because of its viscosity, tar sands bitumen cannot be recovered by conventional drilling and flowing well methods.

Even after recovery, the bitumen's high carbon-to-hydrogen ratio means that it must undergo further processing before it can be refined – either through carbon removal by coking, or hydrogen addition by hydro-cracking. Because of the extra steps required, oil produced from tar sands is often referred to as synthetic oil.

Interest in the tar sands as a source of petroleum began in the late nineteenth century when the federal government initiated research on the technical problems in recovering the oil.

After the First World War, the Research Council of Alberta and the University of Alberta took up the task of tar sands development. In 1929, the Research Council built an experimental separation plant near Fort McMurray. The project was headed by Karl Clark of the University of Alberta. In the council's 1930 annual report, Clark reported that the plant had operated successfully.

However, little further work was done until after the Second World War, when the Alberta government built a plant at Bitumount. By 1950, the work at Bitumount had demonstrated the technical feasibility of the oil sands separation process.

In 1953, the Great Canadian Oil Sands consortium, now reorganized as Suncor, was formed. After protracted negotiations between the provincial and federal governments, and considerable experimental work, the first large-scale commercial plant was built north of Fort McMurray between 1964 and 1968. Since it opened, the Suncor plant has doubled its production capacity to 9857 cubic metres a day.

The sharp rise in oil prices in the early 1970s led to plans for a second and far more ambitious oil sands project – Syncrude, developed by a consortium of oil companies and the federal and Alberta governments. The Syncrude agreement was signed in 1974 and the plant was completed in 1978. Although production is about twice that of Suncor, Syncrude's price tag of $2.2 billion made it 10 times more expensive.

The size of the project alone marks it as a major engineering achievement. The plant is both a mine and a refinery. The mine's total overburden and volume of ore make it one of the biggest open pit operations in the world – the biggest in terms of the amount of ore processed. Also the biggest in the world are the project's fluid coking units.

Locating an operation on top of muskeg and oil sands and building a plant that could mine and process sticky, abrasive tar sand in temperatures ranging from $-50°C$ in winter to $40°C$ in summer required extremely innovative engineering. Prefabrication and preassembly techniques were used to overcome the problem of cold temperatures which confined construction to the summer months. Because of the low bearing-strength and susceptibility to frost-heaving of muskeg, three million cubic metres of it were removed and replaced with granular fill before the foundations could be laid.

Even after the muskeg was removed, some of the foundations extended to the unstable, underlying oil sand. Construction of these foundations was based on a new model devised to predict the settlement of loaded foundations on oil sand over time. Special air-cooled foundations were developed for the hot bitumen tanks to prevent the oil sands in which they were placed from weakening and the tanks from rupturing or tipping.

Additional development work included conveyor-belt materials and idler components (for wheels, gears, and pulleys) capable of operating under low temperatures and heavy loads, special greases for the mining machinery, and bucket wheel teeth that could withstand high-impact loads in conditions of extreme cold.

All in all, the Syncrude project required innovative design and engineering on a massive scale. The overall electrical power needs were, for example, equivalent to those of a city of 300 000 people. In addition to setting up the mining, extraction, and refining operations, the project had to create a town of 1800 single-family dwellings and 3500 apartments.

A towering, bucket-wheel excavator strips oil sand at the open pit mine at Fort McMurray, Alberta. The oil sand is dynamited before excavation to facilitate digging and to prolong the life of digging teeth and buckets. After excavation, the oil sand is transferred to the extraction plant by a system of conveyor belts. (Canadian Government Photo Centre 76-1058)

Early Exploration in the Far North
Petroleum development began in the Far North in 1920, when Imperial Oil Limited brought in a well about 160 kilometres south of the Arctic Circle at Norman (later Norman Wells) on the Mackenzie River. As there was no economical way of shipping the oil south, the well's output was used to supply modest regional needs for fuel for heaters, lighting plants, and aircraft.

In the mid 1930s, the development of pitchblende deposits on Great Slave Lake and gold mines at Yellowknife increased the demand for fuel. Imperial Oil built a full-fledged refinery at Norman Wells and a 13.5-kilometre pipeline around the rapids so that fuel could be loaded onto boats. Although modest by most standards, the refinery's production of 500 barrels (79.5 cubic metres) a day was nevertheless of great

importance in providing a local source of fuel for northern exploration and economic development.

The Second World War created a rapid escalation in the demand for oil from Norman Wells. The United States Army decided to build a highway from Dawson Creek, British Columbia, to Alaska via the Yukon. The highway was needed to open up supply routes to defend Alaska against the perceived Japanese threat. The plan included the expansion of oil production at Norman Wells, and a 952-kilometre pipeline to deliver 13 500 litres of crude oil per day from Norman Wells to a new refinery at Whitehorse.

The Canol (Canadian Oil) contract, a complex deal involving both Canadian and U.S. governments, Imperial Oil, and others, was signed in May 1942 and, by mid-July, the first new well at Norman Wells began production. By 1945, 636 cubic metres a day were flowing from 64 wells. The Alaska Highway, it turned out, was never needed for defence, though it did provide facilities and training for postwar expansion in the North.

Northern exploration and drilling activity during the 1950s and 1960s proceeded sporadically. In 1957, a Calgary company, Western Minerals, drilled the first well north of the Arctic Circle at Eagle Plains, 800 kilometres from Whitehorse. During the winter of 1961–62, the first well in the Arctic Islands was drilled at Winter Harbour on Melville Island and two more test wells were later drilled at the same location.

Although all of these early sites were eventually abandoned, they did demonstrate the feasibility of drilling in the Arctic, and they set the stage for some of the most impressive feats of twentieth-century petroleum and marine engineering.

Artificial Islands
Early exploration had shown that the geology of the Arctic Islands and the surrounding seabed appeared extremely promising for petroleum deposits. In addition, sea access to the project sites would lower transportation, exploration, and production costs.

In the early 1970s, exploration from artificial islands began in the Canadian Beaufort Sea. The first artificial islands were made with sand fill – using either hydraulic fill dredged from deeper waters or solid fill dropped through land-fast ice in shallow waters.

The method was time-consuming and required large quantities of fill. In addition, the size of the islands was limited by the fact that the entire structure had to be completed in one short summer season.

One solution was to use caissons to retain the sand fill. This reduced the quantities of sand needed and the loss of fill from erosion in the wave zone.

The Tarsuit Caissons, built by Swan Wooster and Dome Petroleum engineers, represented the first use of caissons for artificial island construction. Their success demonstrated that such structures could be deployed in the short Arctic summer and that they were effective in overcoming the problems posed by wave erosion and ice loads.

In size alone, the Tarsuit Caissons posed significant engineering challenges. Each of the four floating concrete caissons weighed 5800 tonnes and measured 70 metres in length, 15 metres in width, and 11 metres in height. The caissons were built in Vancouver and floated on a specially built submersible barge to the Beaufort Sea.

Once the caissons were on site, they were joined to form a square over a base or berm prepared from dredged sand. The berm extended from the seabed to six metres below the waterline. When they were in place over the berm, the caissons were filled with

The success of the approach used in the Tarsuit Caissons for constructing artificial islands as oil drilling platforms led to a number of similar projects in the Canadian Beaufort Sea. The ring-like structure under construction in this 1985 photograph is Esso Resources' Caisson Retained Island Project. The ship in the foreground served as a floating industrial base during construction. (Ranson Photographers Ltd.)

Molikpaq, the mobile Arctic caisson, seen here in the Canadian Beaufort Sea, is designed and constructed for year-round drilling in ice-filled Arctic waters. For drilling operations, the unit is towed to a location and then lowered using sea water as ballast onto a pre-dredged subsurface foundation. When drilling is complete, the caisson is easily refloated and towed to another location. Drilling equipment and housing for 100 are built onto the unit's main deck. (Gulf Canada, Ranson Photographers Ltd.)

dredged sand to provide a stable work platform of sufficient mass to resist the huge ice forces in the Arctic.

As the first offshore structure installed in the hostile ice environment of the Beaufort Sea, and the first in which the caisson concept was used for artificial island construction, the Tarsuit Caissons project was surrounded with a good deal of uncertainty. To ensure that the platform could withstand the enormous stresses and abrasive forces of moving ice, it was built of a specially formulated, lightweight, super-plasticized, high-strength concrete with approximately four times more reinforcing steel than normal.

The Tarsuit Caissons demonstrated conclusively that caissons can provide effective platforms for drilling in the Arctic. The project attracted international acclaim and, by 1984, three more caisson systems had been built in the Beaufort Sea.

Another notable platform, Gulf Canada's Molikpaq, embodies a second original Canadian approach to offshore Arctic drilling. In terms of steel weight, the Gulf Molikpaq is one of the largest exploration structures in the world. Capable of drilling in water up to 40 metres deep, it is an unprotected structure with no shallow surrounding underwater berm to ground out the ice. Instead, it is built sturdily enough to withstand direct hits from big ice floes.

Because of the success of the Molikpaq, Canadian engineers were called in to design similar structures in the Alaskan Beaufort Sea, in Sakhalin in the Soviet Union, and in the Barents Sea.

The most recent and perhaps the most creative Canadian approach to Arctic offshore drilling is using the ice itself to build floating exploration platforms. Early efforts to flood existing ice with water proved too slow – the ice thickness increased by only a few centimetres a day. The important breakthrough came with the development of spraying techniques that allowed the ice to build up at the rate of 0.5 metres a day or more.

After experiments in using the spraying method to build ice islands in shallow areas of the Canadian Beaufort Sea, a test island was built in the Alaskan area. The first well drilled from a spray ice-pad was sunk in Alaska in 1986 in water about nine metres deep.

Tarsuit, Molikpaq, and the ice islands are three of the most important milestones in Canadian offshore engineering. They represent engineering in extreme conditions and at the very frontiers of knowledge. Instruments at Tarsuit and Molikpaq are being used to monitor and measure the load on these platforms. The information will be extremely useful in developing future drilling platforms and other structures in the Arctic.

The Ocean Ranger

Whether in the Arctic, the North Sea, or the Atlantic, offshore drilling involves the employment of extremely complex systems in a hostile, hazardous, and unpredictable environment. The danger of these operations was borne out when, at 1:30 a.m. on Monday, February 15, 1982, the semisubmersible, self-propelled drilling rig, the Ocean Ranger, radioed from its drilling location on the Continental Shelf 170 nautical miles east of St. John's, Newfoundland, that its crew was going to the lifeboat stations and that the rig was being evacuated. The entire 84-man crew perished in the frigid waters of the Grand Banks. The accident was a major tragedy for Newfoundland: of the 69 Canadian crew members, 56 were residents of that province.

A Royal Commission of Inquiry was called "to examine not only the cause of the loss, but also areas of vulnerability within which lay the potential for this disaster and the seeds for future ones."[40] The commission found that both the design and the management of the rig had contributed to the disaster.

On the Ocean Ranger's design, the commissioners wrote:

Various aspects of the Ocean Ranger's design made the rig more vulnerable than it should have been: the exposed location of the ballast control room; the weakness of the portlight; the lack of protection against sea water of the mimic panel; the lack of an adequate manual control system in the control room and the vulnerability of the chain lockers.[41]

The commissioners found a number of flaws in the management of the rig as well.

Aspects of the management system were also contributory factors: lack of proper procedures for emergencies, lack of manuals and technical information relating to the ballast control console and lack of adequate training programs for key personnel.[42]

The accident, the commissioners concluded, resulted from a combination of human error and of shortcomings in design and management.

Each event and action which contributed to the loss of the Ocean Ranger was either the result of design deficiencies or was crew-initiated. The disaster could have been avoided by relatively minor modifications to the design of the rig and its systems and it should, in any event, have been prevented by competent and informed action by those on board. Because of inadequate training and lack of manuals and technical information, the crew failed to interrupt the fatal chain of events which led to the eventual loss of the Ocean Ranger. It is, nevertheless, the essence of good design to reduce the possibility of human error and of good management to ensure that employees receive training adequate to their responsibilities.[43]

The legacy of the Ocean Ranger tragedy has been a number of research, engineering, and legislative measures all designed to reduce the risks of offshore drilling-rig accidents and to prepare crews to take proper safety actions when they occur. For example, at the time of the Royal Commission, it was noted that there had been very little progress in developing effective evacuation systems for offshore rigs. Lifeboats were difficult to launch and there was a significant danger that they would smash into the rig after launching. Since then, approximately 50 systems have been tested in Canada for use on offshore rigs.

Since the 1960s, highrise buildings have dramatically reshaped the downtown cores of many Canadian cities. This photograph of a highrise under construction in Calgary, Alberta, in 1970 shows many features typical of highrise building: flying cranes, concrete and reusable concrete forms, and a reliance on off-site manufactured building components. (Canadian Government Photo Centre 70-1672)

Improvements in engineering, management, safety standards, and equipment cannot entirely prevent another accident. The Ocean Ranger disaster does, however, underscore the great responsibility that today's engineers have to reduce the risks associated with the powerful and complex systems they create.

Highrise Monuments to Concrete and Steel

Since the late 1960s, economic and social pressures have created a Canadian population which is increasingly urban and white collar. As a result, there has been a stiff demand for residences and offices in urban centres, which has been met by a sometimes frenetic construction of highrise towers. These have now become the dominant architectural and engineering form in Canadian cities.

In contrast with its structural-steel predecessor, the modern highrise is a monument to concrete. Higher-strength concretes capable of carrying heavier loads have made highrise construction possible, and improvements in concrete construction techniques have allowed highrise towers to be built quickly and economically.

Canadian engineers pioneered one of the most significant of these techniques – the flying form. In this system, a crane raises the forms up the building to produce layer after layer of concrete floor. The use of standardized, interchangeable components has been a major reason for the proliferation of highrises.

Many of these towers are simply bland utilitarian structures. At times, urban planning considerations have been ignored and highrises have created problems of overcrowding and traffic congestion. Nevertheless, they are generally far more pleasant to live in and work in than were their predecessors. In some instances, they are exceptionally fine buildings.

Completed in 1970, the West Coast Transmission Building is one of Canada's most striking modern highrises. Because Vancouver is located in an earthquake zone, the West Coast Transmission Building was designed to minimize damage and loss of life during an earthquake. The cables running out from the core at the top of the building are an integral part of Dr. Bogue Babicki's innovative design and engineering. Suspending the floors from the central core, rather than rigidly attaching them to the framework as in conventional highrise construction, allows the building to move more freely and with less damage during an earthquake. (Bogue Babicki Associates Inc.)

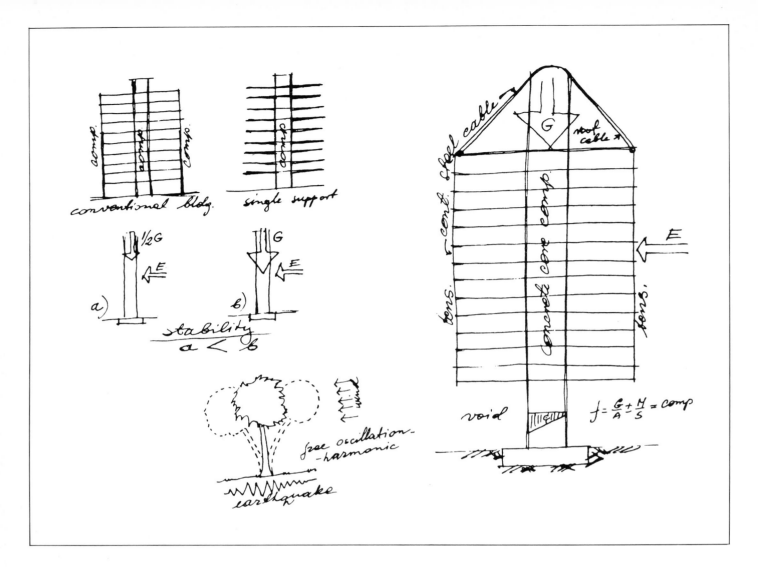

Engineering to Withstand Earthquakes

The West Coast Transmission Building in Vancouver is one of these. With its hilly terrain and splendid harbour overlooked by snowcapped mountains, Vancouver is one of Canada's most beautiful cities. The West Coast Transmission Building, built by the Vancouver engineer Dr. Bogue Babicki, is designed to blend harmoniously with the city's spectacular setting. The result is a building that is striking, elegant, and functional.

Part of the reason for the unusual design is Dr. Babicki's original approach to engineering a building that can withstand earthquakes.

During earthquakes and their aftershocks, much of the damage to buildings and injury to occupants is caused by the misalignment of walls and floors. With the vertical and horizontal support systems joined rigidly together, deflection in one causes problems in the other.

Babicki had observed that trees are rarely damaged in earthquakes. Although the trunk of the tree, which is a strong, flexible central column or shaft, may sway violently, the lighter branches that radiate out from this column can move freely without causing damage.

Babicki decided to apply the lessons he had learned from studying the structure of trees in designing the West Coast Transmission Building so that it would be earthquake-resistant. Rather than rigidly interconnected parts, the building has two related but independent systems.

These sketches by Dr. Bogue Babicki illustrate the principles he applied in the innovative, earthquake-resistant design of the West Coast Transmission Building in Vancouver. Dr. Babicki observed that trees are rarely damaged during an earthquake. The trunk of the tree – a strong, flexible central column or shaft – may sway violently, but the lighter branches that radiate out from the trunk move without causing damage. These ideas were first applied in the West Coast Transmission Building and subsequently in a number of other buildings designed by Dr. Babicki. (Bogue Babicki Associates Inc.)

In 1980, Dr. Bogue Babicki designed the Monreale Complex in Cagliari, Sardinia, Italy. These ten cable-suspended office towers are a further development of the principles used in Vancouver's West Coast Transmission Building. The ten buildings form a curved line, enclosing the largest commercial and residential complex on the earthquake-prone island. The photograph above, of the complex under construction, illustrates the sophisticated approach used to manage the project. Much of the work on the various buildings was scheduled sequentially. This reduced the costs of expensive equipment through the elimination of unnecessary duplication. (Bogue Babicki Associates Inc.)

At the centre of the building, there is a square, hollow, concrete core which contains elevators and other services. This is analogous to the trunk of a tree. Floors are suspended from the top of this core. Like the branches of a tree, they are able to move independently during an earthquake.

From a structural point of view, only the core needed contact with the ground. Babicki exploited this fact in creating a design in which the floors remain suspended in the air. Unlike other highrises, where the only good view is from the upper offices, the West Coast Transmission Building permits a view and is attractive to look at from virtually any vantage point. Even from ground level, for example, one can look through the building to the harbour and mountains beyond.

The powerful impact of the building's design was demonstrated when the author mentioned to a Vancouver cab driver that he was in the city doing research for a book on engineering in Canada. The driver asked if he could go a few blocks out of the way to point out something that was worth writing about. In a few minutes, he stopped and gestured toward the West Coast Transmission Building. "I don't know what holds the floors up," he said. "They must be hanging from something. That's the only building like it in Vancouver – maybe in the world."

Babicki's work on the West Coast Transmission Building shows that an innovative engineer can make a safe, functional, and aesthetically pleasing highrise. His approach is not yet popular or even widely known in Canada. However, in earthquake-prone Sardinia, Italy, Babicki has used the same principles in designing ten 11-story, cable-suspended office towers which are part of a large residential and commercial complex.

Designing the world's tallest free-standing structure, the CN Tower in Toronto, presented unprecedented challenges and required extensive testing. Here a 1:500 aero-elastic model of the CN Tower is tested in the boundary layer wind tunnel at the University of Western Ontario. (Boundary Layer Wind Tunnel Laboratory, University of Western Ontario)

The CN Tower: A Multidisciplinary Effort

For engineers in the nineteenth century, iron and steel were exciting new materials. The Eiffel Tower stands as a monument to the exuberant exploration of the structural possibilities of these materials. In a similar manner, Toronto's CN Tower pushes concrete, the dominant engineering material of the latter part of the twentieth century, to its structural limits.

Despite widespread criticism at the time of construction, both towers have become symbols of their age and major tourist attractions. Since its opening in 1976, for example, more than 15 million people have visited the CN Tower.

Rising to a height of 553 metres, the tower is the world's tallest freestanding structure. It is also a utilitarian structure – a communications tower.

During construction of the CN Tower, the crane moved up with the tower and was used to lift equipment and materials from the ground. To produce the graceful taper of the CN Tower, the elaborate slipforms, visible in the photograph around the tower, were reduced in size as they were pushed skyward. (Panda Associates 73101-189)

Raising the brackets and formwork to build the seven-story SkyPod at the 335-metre level of the CN Tower was an extremely delicate operation. Here one sees three of the six separate sections of formwork that were raised in unison, a few centimetres at a time, by means of steel cables pulled by 45 hydraulic jacks. Concrete was then poured into the formwork to make the twelve 10.4 by 10.4-metre triangular brackets that support the observation and communications pod. (Panda Associates 73101-265)

This photograph was taken from a giant
Sikorsky Skycrane helicopter as it lifted
the last section of the antenna to the top
of the 553-metre CN Tower on April 2,
1975. (Foundation Co. of Canada Ltd.)

The need for a high communications tower in Toronto became acute during the late 1960s. Highrise buildings and electromagnetic interference were making it virtually impossible to transmit clear signals from towers in the downtown core. A tower higher than any other existing or planned structure would be able to combine UHF, TV, and FM broadcast transmissions with microwave communications that require a clear line of sight. With the diminishing importance of railroads in the city centre, railway land became available for the tower near Toronto's downtown waterfront.

Preliminary designs were subjected to stringent testing for stability in the University of Western Ontario's Boundary Layer Wind Tunnel, and a final design was selected in 1972.

A foundation exploration program was then begun by engineering consultant Dr. Eli Robinsky to ensure that the foundations would be capable of bearing the tower's extremely high loads. The site itself was part of lands reclaimed from Lake Ontario since the nineteenth century. Four test holes, each 76 centimetres in diameter, were drilled through over nine metres of surface earth and then 27 metres into the grey Dundas shale. Dr. Robinsky took samples for laboratory analysis, and he and a photographer were lowered into the bore holes for direct observation, sampling, and photography.

Following Dr. Robinsky's recommendations for the foundations, all of the overburden and over 56 000 tonnes of shale were removed, bringing the total excavation depth to about 15 metres below the surface. The remaining shale surface was smoothed and cleaned. Then, to ensure that the newly exposed shale surface did not deteriorate, it was covered with burlap and kept wet.

More permanent protection of the shale was provided by a concrete pad 30 centimetres thick. The pad then became the base for the seven-metre-thick concrete foundation. The foundation alone required 7060 cubic metres of concrete, 454 tonnes of reinforcing steel, and another 36 tonnes of thick tensioning cables.

Concrete work on the tower itself presented further challenges. In traditional construction, concrete is poured into static forms. When it sets, the forms are removed. To make the structure higher, the forms are placed higher and additional concrete poured.

The alternative method is to use slip forms, which are pushed slowly and continuously upward by hydraulic jacks as the concrete sets. The slip-form method is quicker and provides a bonded, monolithic structure, rather than a series of jointed portions.

The slip-form method is well-suited to the straight, parallel sides of most concrete structures. However, as the CN Tower rises in height, its Y-shaped cross section becomes smaller and the projecting portions less pronounced. Slip-form designers were able to overcome the problem of the CN Tower's continuously tapering sides by designing forms that grew smaller and tapered as they rose.

The CN Tower also required very high-strength concrete. The 6000-psi concrete that was developed exceeded original design specifications. Once pouring started, it continued around the clock for eight months. On some days, the slip forms climbed as much as six metres.

Throughout the construction, post-tensioning cables as well as more conventional reinforcing were used. The cables produced a structure under permanent tension, no matter how heavily it is loaded by high winds. With permanent tension, the tower should remain free of tiny cracks which would allow water and atmospheric pollution to enter and eventually destroy the reinforcing steel and thereby the strength and stability of the tower. The post-tensioning cables will thus add greatly to the tower's life span.

Steps were taken to counter the serious problem of falling ice experienced with a similar – though smaller – tower in the Soviet Union. Blocks of ice weighing as much as 136 kilograms fall from that tower and land as far as half a kilometre away.

One approach to this problem was to minimize the number of protrusions in the overall design. Another was the installation of electric heaters where icicles might form on the cladding. The most serious ice problem was, however, created by the large parabolic dishes used for microwave transmission and reception. The entire area in which the dishes are located was covered with a flexible ring of air-inflated radome. The radome's teflon coating and shape cause any accumulation of ice to fall harmlessly as small particles of powdery snow.

Excavation for construction of the CN Tower began in February 1973. A year later, on February 22, 1974, the top of the slip-form concrete was reached at a height of 446 metres. An additional five metres of concrete were added to support the steel transmission mast which itself rose another 102 metres. The top of the mast was lifted into position by the world's largest helicopter.

The conception, design, and construction of the CN Tower reflect the complexity of modern engineering and the degree to which engineers now work in multi-disciplinary teams rather than as individuals. Specialists involved in the CN project included architects; design consultants; structural, mechanical, electrical, and broadcast engineers; construction managers; concrete and lighting experts; soil analysts; and wind tunnel and vibration analysis experts.

The CN Tower required the best efforts of leading engineers and experts from a wide variety of fields. The fact that this twentieth-century equivalent of the Eiffel Tower now stands in a Canadian city is an indication of the level of maturity Canadian engineering has achieved.

Engineering in a Society Increasingly Concerned with Equality

While in the midst of change, it is often difficult to determine what is significant in the past; however, with the passage of time, historians should develop a clearer perspective on the tumultuous decades that mark the end of this 100 years. Clearly, the better part of this period evidences a general social trend toward greater equality and recognition of self-worth. With respect to engineering, this was especially true in the recognition of language rights, the role of women, and rights of the physically disabled.

Language, Sourcing, and Quebec Engineering

French-speaking Canadians have a long-standing, strong sense of pride in and identity with their language and culture. But, by the 1960s, many French-speaking Canadians were increasingly frustrated. In many government and business circles in Canada, English was the dominant, often the only, language at middle and higher levels of operation. Many French-speaking professionals saw their own language as an impediment to well-deserved promotion and opportunity. French Canadians in Quebec, where French is the first language of over 82 per cent of the population, felt that they were denied full participation in the political and economic life of their own province.

Growing unrest and awareness of the extent of language inequality led the federal government to appoint a Royal Commission on Bilingualism and Biculturalism in 1963. The commission's preliminary report in 1965 and its six volumes of final reports between 1967 and 1970 set the stage for a pronounced shift in attitudes, legislation, and practices.

Pierre Elliott Trudeau, who became Prime Minister of Canada in 1968 and remained in power during much of the period of most rapid change, played a central role.

Bombardier is one of many Quebec-based engineering firms that have grown rapidly in recent years. Bombardier manufactures these push-pull commuter cars for the Metro-North Commuter Railroad, a division of the Metropolitan Transportation Authority of New York. (Bombardier Inc.)

Trudeau was strongly committed to the idea that a federal system of government was workable in Canada and that it was possible to make the country one in which both Anglophones and Francophones could feel comfortable. Such a commitment to abstract and philosophical notions is rare in Canadian history. Despite the considerable controversy and misunderstanding they caused, Trudeau clung tenaciously to the ideals of federalism and language equality.

The Official Languages Act of 1969 proclaimed that English and French were Canada's official languages. The act opened up new employment opportunities for French Canadians. Even more importantly, it provided official recognition of the legitimacy and equality of the French language in Canada. The recognition of legitimacy added greatly to the confidence of French-speaking engineers and strengthened even further their rapidly accelerating record of achievement.

Hydro-Québec was instrumental in fostering the growth of Quebec-based engineering firms. During its period of rapid growth, Hydro-Québec was involved in some of the world's biggest and most sophisticated hydro-electric engineering and construction jobs. Ambitious projects such as the James Bay hydro-electric development required enormous engineering input. Hydro-Québec faced a crucial choice; it could do as Ontario Hydro had and rely heavily on an expanding in-house engineering capability or it could draw largely from private-sector consulting firms. Hydro-Québec chose the latter and in so doing gave private-sector firms the opportunity to play leading roles on the world's most sophisticated and challenging hydro-electric projects. Companies such as SNC, Monenco, and Lavalin were later able to expand rapidly through worldwide marketing of the expertise and reputation they had acquired on Hydro-Québec projects.

Bombardier provides another example of the confident and exuberant growth of Quebec engineering. It acquired the rights to Montreal's subway technology, with its rubber-tired cars, from the original French manufacturer, and won the contract to supply new cars when the system was expanding prior to the 1976 Summer Olympics. Using this contract as a launching pad, the world's foremost snowmobile manufacturer (see chapter 4) was able to become a major player in the public transit sector. Today, Bombardier subway cars, labelled "Fabriqué par Bombardier Inc., La Pocatière, Canada," can be seen not only in Montreal, but also in the sophisticated transit system in Mexico City. The company has now diversified into the aircraft industry with the recent acquisition of Canadair.

Women and Engineering

A more recent trend is the increasing importance of women in engineering. Women now represent 16 per cent of the engineering student population and 2 per cent of the professional population. Although these figures are lower than those for other professions and do not come close to the percentage of professional women in the workforce, it is an improvement.

During the nineteenth century, engineering was seen as inappropriate for women. Universities were often reluctant to admit women to engineering programs and women were reluctant to apply. It was not until 1927 that the University of Toronto graduated its first woman electrical engineer, Elsie MacGill. Earlier in the same decade, Llewellyn May Jones became the first woman engineering graduate from King's College in Halifax. For many women the thought of engineering was ruled out by family, guidance counsellors or teachers, and a society that generally viewed science and mathematics as unsuitable for women. Consequently, many women university applicants did not have the necessary prerequisites to study engineering.

Women who wanted to pursue a science-related career found that professions such as medicine, dentistry, pharmacy, and chemistry offered more status and better pay. In addition, these professions made it easier to leave and re-enter the work force.

Before World War II, women in Canada earned 45 per cent of all bachelor degrees and 13 per cent of all doctorates. Following the war, veterans overtaxed university facilities and few women gained admittance to engineering schools. The percentage of women studying at university did not return to prewar levels until the 1970s. As more women entered the profession, some employers, such as Ontario Hydro and Hydro-Québec, became known for their willingness to give them a fair chance. Women like Claudette MacKay-Lassonde and Danielle Zaikoff demonstrated that women could rise within the executive ranks and not forever be relegated to the lower echelons. In 1975, Danielle Zaikoff became the first women president of the Ordre des Ingénieurs du Québec then of the Canadian Council of Professional Engineers.

It is likely that, as clearer historical patterns emerge, the 1970s will be seen as the beginning of a trend toward more Canadian-trained women becoming engineers and the 1980s as the start of a trend toward more women in major management positions. Another factor contributing to this pattern is the growth of organizations such as Women in Science and Engineering (WISE).

Much of the credit for the changing role of women in engineering is owed to those pioneers who entered the profession, stayed with it, and rose to positions of authority. Women, such as Claudette MacKay-Lassonde (first woman president of the Association of Professional Engineers of Ontario), Dr. Ursula Franklin, Dr. Dormer Ellis, Maureen Lofthouse, Micheline Bouchard, Denise Therrien-Bolullo, Michele Thibodeau DeGuire, and Michele Boucher, are Canadian pioneers.

Rehabilitation: The Engineering of Compassion

After the Second World War, returning troopships brought the seriously wounded and disabled back to veterans' hospitals and homes in Canada. Among the injured veterans were many amputees. Their plight helped trigger a series of important Canadian engineering advances that have made Canada a world leader in engineering to help people with severe physical disabilities.

Artificial Limbs

Today, James Foort and his colleagues at the Medical Engineering Research Unit of Shaughnessey Hospital in Vancouver are working in the solid tradition of rehabilitation engineering born of the Second World War. They are using modern computer technology to solve one of the most critical problems in prosthetics – ensuring a proper fit for an artificial limb.

Prostheses are difficult to fit because in size, configuration, and extent of damage, each stump is unique. A poorly fitting artificial limb will be of only limited use and can be extremely painful. It can press on nerves, reduce blood circulation by improper and excessive pressure, and lead to abrasion and chafing.

Traditionally, the fitting of artificial limbs was a time-consuming, trial-and-error process. It was largely dependent on the skill, experience, and patience of the prosthetist.

James Foort and his colleagues have developed an ingenious CAD/CAM (computer-assisted design/computer-assisted manufacturing) system which reduces the time involved in this process. In Foort's system, two cameras work simultaneously from different vantage points to create a three-dimensional map of the stump. The map is digitized and stored in computer-usable form. From this record, a computer-controlled milling machine carves a full-scale model of the stump which is then used to make the plastic socket that connects the stump to the artificial limb.

After the socket is made, the next step is to fine tune it so that it fits as well and as comfortably as possible. Pressure on sensitive points must be minimized; other points will need more pressure; and each adjustment in one area has an effect in others.

Under traditional techniques, each adjustment required physical modifications to the socket or the manufacture of an entirely new one. Foort's system, however, uses computer technology to minimize these time-consuming and expensive changes.

As the patient describes to the prosthetist where the fit is poor, the digitized stump map, complete with contour lines, is viewed on a television screen. By using a computer "mouse" or light pencil, the prosthetist modifies the image and then rotates it to view the implications of these changes from any desired angle. When the fit seems right, the image is recorded and used to mill another model which is then used to make a new socket.

Foort's new system helps prosthetists create better fitting artificial limbs far more quickly and cost-effectively. It also frees them from routine work, allowing them to treat many more patients. It represents the application of leading-edge technology from a number of engineering fields – computer-assisted design, digital control of machine tools, and the manufacture of plastics and other synthetic materials – to give better treatment to the physically disabled.

Engineering for Thalidomide Children

In the early 1960s, the synthetic drug thalidomide became popular as an anti-nausea remedy for pregnant women. Thousands of deformed children were born around the world before it was discovered that the drug caused birth defects and it was taken off the market.

In Canada, a number of programs were established to deal with the special problems of thalidomide-deformed children. In Toronto, the Prosthetic Research and Training Unit at the Ontario Crippled Children's Centre was created in 1963, and projects were begun in Winnipeg under the Manitoba Sanitorium Board and in Montreal at the Rehabilitation Institute of Montreal.

All these institutions worked together with special funding from Health and Welfare Canada. Although the programs were created to respond to the special needs of the 120 thalidomide children in Canada, the research was aimed at prosthetics to help all of the physically disabled.

Early work involved the development of electro-mechanical upper limbs. These could be activated either by mechanical switches moved by stumps or by myo-electric switches responding to electrical impulses from muscle remnants. Research and design on the myo-electric systems were carried out under the direction of Professor Bob Scott at the University of New Brunswick.

Additional research and design were aimed at meeting the special seating needs of handicapped people who suffer from such diseases as cerebral palsy and spina bifida. Modular systems were developed which allowed seats to be tailored to individual need. As well as providing seating that was comfortable, functional, and that impeded the progress of the disease, the modular design removed the necessity of manufacturing a seat for each individual and meant that costs could be kept low.

The researchers also focussed on means to improve the mobility and communications of disabled people. A wide variety of control mechanisms, carts, signalling devices, and specialized vehicles were produced, and computer systems that permit the use of conventional computers and software were developed.

Many of the fruits of the research and development program were made available through the Variety Village Electrolimb Production Centre, a facility that designs, manufactures, and markets prosthetic and orthotic products for children and young adults.

Canadian society responded to the thalidomide tragedy by funding an extensive program of scientific and engineering research. The products developed as a result of this research have benefited not only the thalidomide children but many other disabled people as well.

The Orion II
For those confined to wheelchairs, transportation with dignity is a recurring problem. Conventional cars and buses cannot accommodate wheelchairs, so in the past the physically handicapped have had to rely on vans that are specially equipped with hydraulic lifts or elevators.

The Orion II is a bus that solves the problem of transportation for those confined to wheelchairs, and it does it elegantly, economically, and reliably. Rather than

requiring special equipment and assistance, disabled people are able to get on and off the bus unassisted.

Manufactured by Ontario Bus Industries of Mississauga, Ontario, the Orion II is a product of collaboration between engineers and one of Canada's foremost industrial designers, Claude Gidman.

The Orion II lowers itself on air suspension to within five centimetres of the curb or 10 centimetres of the ground. Individuals can then wheel themselves on or off the bus on flip-down ramps.

The secret of the bus's ability to lower the floor level so close to the ground is the placement of its main structural support in the roof frame. In effect, the rest of the bus is suspended from this frame, which allows the flat, thin, uncluttered floor to settle close to the ground for loading and unloading.

The Orion II is sturdily built with a 20-year life expectancy. It is also designed for rapid servicing. The front power-train module, which consists of the engine and cooling system, the transmission, the front-wheel-drive assembly, the suspension, and the steering, can be removed and replaced within two hours. In addition to its ability to accommodate conventional seats and wheelchairs, the Orion II can be converted into a multipassenger ambulance.

Like the work by James Foort's group at the Medical Engineering Research Unit of Shaughnessey Hospital and the the remarkable devices developed in the wake of the thalidomide tragedy, the Orion II is an excellent example of the way in which Canadian engineering is helping the physically disabled.

Pushing Back the Frontiers

No technological achievement of the 1960s, 1970s, and 1980s has received greater public attention than the space program. The first halting steps to explore space must certainly rank as one of the outstanding achievements of our age. And Canadians are justly proud of the Canadarm which has been used so successfully in the United States Space Shuttle program.

Nevertheless, Canadian achievements in another area of exploration – the seabed – have perhaps received far less attention than they deserve. The seabed has recently become a centre of vigorous activity largely because of dramatic improvements in underwater craft, a field in which Canada is a world contender. In fact, the world's largest manufacturer of unmanned civilian submersibles is International Submarine Engineering Ltd. (ISE) of Port Moody, British Columbia.

The founder of ISE is James McFarlane, an ex-naval commander with degrees in engineering and naval architecture from the University of New Brunswick and the Massachusetts Institute of Technology. McFarlane's experience in working on Canada's Oberon submarines in England provided him with the background to start ISE.

Submersible craft manufactured by ISE are used for a wide range of purposes, including mining seabed deposits, capping undersea oil well blowouts, seeking sunken treasure, and surveying and mapping under water and ice. ISE components were employed, for example, in locating and mapping the wreckage of the downed Air India passenger jet on the bottom of the Irish Sea in 1985.

ISE's submersibles, plus other striking projects and achievements such as the CN Tower, offshore Arctic drilling, and innovative devices and vehicles for the disabled are just a few of the areas in which Canadian engineering has assumed a leading role in recent decades. These examples of engineering innovation may well point the way to the future.

Canada's engineers have long been recognized as world leaders in areas of traditional Canadian strength – hydro-electric power projects, mines and smelters, and

In 1982, Ontario Bus Industries of Mississauga, Ontario, began work on the prototype of the Orion II, a low-floor bus without steps designed to accommodate disabled and elderly passengers. The Orion II won an Award of Excellence in 1985 for industrial design. An air-bag suspension lowers the unit to curb height at the front door and close to ground level at the rear door. The basic unit can also be configured for fixed route transit, feeder routes to rail transit, suburban bus service, airport shuttle service, and ambulance service. Much of the company's production is exported. (Lorna MacPherson, OC Transpo)

road and railroad construction. Quite naturally, these are also the areas in which Canadian engineering firms have successfully marketed their expertise in other countries.

But foreign competition is making it increasingly difficult for Canada to rely on resource-based industries and engineering for its economic health and prosperity. Moreover, modern communications and transportation are creating a single world economy. Because of this trend, Canadian engineers can expect to compete increasingly in world markets.

Throughout its history, Canadian engineering has successfully responded to new challenges. From John MacTaggart's development of new techniques required by the special conditions of the country during construction of the Rideau Canal in the early part of the

nineteenth century to the ISE's leadership in submersibles today, the ability of Canadian engineers to adapt technology creatively and innovatively has been their single most important achievement.

The economic well-being of Canada in the face of increasingly fierce international competition may well depend on the ability of our engineers to continue developing and strengthening this tradition.

The Canadair CL-215 is the only commercial-sized aircraft specifically designed for water bombing. To be cost-competitive with older aircraft converted to water bombing at little expense, the design was based on existing technology. Here a CL-215, operated by France's Sécurité Civile, blankets a forest fire in Corsica. The CL-215 can scoop 5350 litres in 10 seconds and make over 200 separate attacks on fires in one day. There are now over 100 in service in seven countries. A turbine version is planned to replace the existing design, originally conceived in 1967. (Canadair Photographic Services)

The Canadian engineering firm SNC was the construction management of the building and dome for the Canada-France-Hawaii Telescope in 1974. Shown here before completion, the facility is located 4300 metres above sea level atop Hawaii's Mauna Kea Mountain. High-altitude projects pose a number of challenges: materials and equipment must be transported to the site; the lower oxygen content of the air can reduce worker productivity and lower the power output of internal combustion engines. (The SNC Group)

Another high-altitude project by SNC was this cement plant in the Ecuadorean Andes in the 1970s. (The SNC Group)

The Tintaya copper plant was built by The SNC Group at 4100 metres in the Peruvian Andes and opened in March 1985. (The SNC Group)

Bombardier was the official supplier of snowmobiles and snow-grooming equipment for the 1984 Winter Olympic Games held at Sarajevo, Yugoslavia. Utility snowmobiles in the foreground were used for transportation of officials, judges, rescue teams, and journalists. The large tracked vehicles in the background were used for grooming alpine downhill, slalom, and giant slalom ski runs. (Bombardier Inc.)

Canadian consulting engineers have helped developing countries establish a basic industrial infrastructure. The construction of this ductile iron foundry at Rouiba, Algeria, by The SNC Group is a good example. (The SNC Group)

The Foundation Group built this diversion tunnel of the river by-pass for the Condoroma Dam, Peru. (Foundation Co. Ltd.)

This hydro-electric and irrigation dam was built by Lavalin Inc. on the Magat River in the Philippines. Located in a major earthquake zone, the structures are designed to withstand severe seismic loading. (Lavalin Inc.)

The 384-megawatt Temengo hydro-electric development located on the upper Perak River in Malaysia was completed in 1978 by Lavalin Inc. (Lavalin Inc.)

In 1976, the Association of Consulting Engineers of Canada and *Canadian Consulting Engineer* magazine presented an Award of Excellence to The SNC Group for the design of Idikki Dam in India. A double-curvature, parabolic concrete arch rising 170 metres above the lowest bedrock level, Idikki Dam is the highest dam of its type in mainland Asia.

The Cinkur zinc complex near Karyseri, Turkey, was completed in June 1976 by The SNC Group. Whereas previously Turkey had imported all its zinc, the new facility produced 40 000 tonnes of the metal annually. (The SNC Group)

A lack of adequate and clean water supplies is a serious problem for many Third World villagers. The PVC pump was developed by the Canadian International Development Corporation in conjunction with a team of engineers from the University of Waterloo. Here the low-maintenance, PVC pump is being used by a woman in Sembilan, Malaysia. (CIDC, Neill McKee)

Canadian Foremost is renowned for its special-purpose vehicles. Here a Canadian Foremost Marauder S.F. is at work in the Saudi Arabian desert for Western Geophysical, another Canadian company. (Western Geophysical)

Sanctuary of the Martyr, Algiers,
Democratic Republic of Algeria, was
realized in collaboration with the
Algerian Presidency and the Youth Detail of
Algeria's "Service National." (Lavalin)

The Challenge of the Second Century

Recognition of Canada's increasing dependence on engineering should be the most important realization to come out of this survey of a century of Canadian engineering. Numerous examples all point to the fact that engineering is a normal part of the everyday physical and intellectual environment of Canadians. Both prosperity and the quality of life in Canada depend heavily on how well engineering is nurtured and used; this poses an enormous challenge, which the engineering profession cannot deal with alone.

The engineering profession has earned a place as one of the foundations of Canadian society and life because it has been so dynamic. Long distances, intimidating topography and geology, extreme temperatures, limited economic resources, and new technologies have posed enormous engineering problems. Success in tackling a steady stream of problems has given Canadian engineers their often remarkable energy, confidence, and reputation for a willingness to confront the seemingly impossible.

Although there is no single reason for Canada's enviable reputation as an engineering nation, one stands out: a tradition of mutually beneficial cooperation between private enterprise and government. Canadians have seen that some tasks are beyond the resources that can be mustered by either government or private enterprise alone and have created mechanisms to combine the resources and strengths of both for the benefit of all. From railroads to hydro-electric projects and the quest for petroleum, joint projects have given Canadians the products and services they want and engineers the experience and reputation that have placed them in such demand worldwide.

Engineering is a cumulative activity; one success sets the stage for another. Conversely, failure to participate sets the stage for continuing exclusion. Faced with a century of engineering success, it is all too easy to forget that those successes were hard won and that the availability of an internationally respected engineering capability cannot be turned on and off like water in a tap. Throughout a century of momentous change, Canadian engineers achieved so much because there was continuity where it was needed most. Education and training were ongoing, as were opportunities for continuing employment and exposure to advanced ideas and research results.

During the past hundred years, Canada has become increasingly dependent on the engineering profession, and will continue to do so in the foreseeable future. This means that, for a healthy country and society, care of the engineering profession is a necessity, as is the fact that this care must be the product of interaction between the profession and society at large. The only option is how well the job is done and even that leaves little room for manoeuvrability. However, it does leave room for imagination. One should remember that, more than anything else, Canadian engineering is a story of imagination and new departures at critical junctures. One should also remember that perceptions of the past and present help to shape the future. Moreover, the one hundredth anniversary of Canada's first engineering society comes at an extremely critical time for Canada. That is why the greatest single challenge facing Canada's engineering profession as it enters its second century is to make itself appreciated and understood; without that, there is little chance that the second century will be as rewarding as the first.

Notes

Chapter 1

1. Robert F. Legget, "The Jones Falls Dam on the Rideau Canal, Ontario, Canada," *The Newcomen Society Transactions*, 31 (1957–58/1958–59): 205.
2. John MacTaggart, *Three Years in Canada: An Account of the Actual State of the Country in 1826-7-8, Comprehending Its Resources, Productions, Improvements, and Capabilities; and Including Sketches of the State of Society, Advice to Emigrants, &c. by John MacTaggart, Civil Engineer, in the Service of the British Government*, 2 vols. (London: Henry Colburn, 1829).
3. MacTaggart, *Three Years*, Vol.1, p. 20.
4. MacTaggart, *Three Years*, Vol.2, pp. 94–96.
5. MacTaggart, *Three Years*, Vol.1, p. 108.
6. MacTaggart, *Three Years*, Vol.2, pp. 162–163.
7. Frances Woodward, "The Influence of the Royal Engineers on the Development of B.C.," *B.C. Studies*, 24 (Winter 1974–75): 27.
8. Original publications by Keefer are quite scarce; more readily accessible is a reprint of some of Keefer's writings cited here. H.V. Nelles, ed., *Philosophy of Railroads and Other Essays by T.C. Keefer* (Toronto: University of Toronto Press, 1972), pp. xxv, 11.
9. Nelles, *Keefer*, p. 97.
10. Nelles, *Keefer*, p. xxii.

Chapter 2

11. *Montreal Gazette* 25 February 1887.
12. P. Alex Peterson, "President's Address," *Transactions Canadian Society of Civil Engineers*, 3, pt. 2 (October to December 1894): 307.
13. "The St. Clair River Tunnel," *The Dominion Illustrated* (10 October 1891): 358.
14. Clarence Hogue, André Bolduc, and Daniel Larouche, *Québec: un siècle d'électricité* (Montréal: Éditions Libre Expression, 1979), p.15.
15. A.J. Lawson, "Generation, Distribution and Measurement of Electricity for Light and Power; Appliances Therefor, and Particulars of Canadian Installations," *Transactions Canadian Society of Civil Engineers*, 4, (1890): 179–240.
16. Fred A. Bowman, "Some Applications of Electric Motors," *Transactions Canadian Society of Civil Engineers*, 8, pt.1 (January–June 1894): 195.

Chapter 3

17. Richard Clippingdale, *Laurier: His Life and World* (Toronto: McGraw-Hill Ryerson, 1979), p. 72.
18. Clippingdale, *Laurier*, p. 72.
19. A. Mitchner, "The Bow River Scheme: CPR's Irrigation Block" in Hugh A. Dempsey, *The CPR West: The Iron Road and the Making of a Nation* (Vancouver: Douglas & McIntyre, 1984), p. 259.
20. M. Murphy, "Concrete as a Substitute for Masonry in Bridge Work," *Transactions Canadian Society of Civil Engineers*, 2 (1888): 80.
21. Murphy, Concrete, pp. 102-103.
22. Walter J. Francis, "Mechanical Canal Locks in Canada," *Transactions Canadian Society of Civil Engineers*, 21 (1906): 77–78.
23. Francis, Mechanical Canal Locks, p. 100.
24. Esther Phyllis Rose, *Frank Barber and His Bridges* (Master's thesis, University of Toronto, 1985), p. 79.
25. Rose, *Frank Barber*, p. 67.
26. C.R. Young, "Bridge Building," *The Engineering Journal*, 20, 6 (June 1937): 492.
27. Rose, *Frank Barber*, p. 1.
28. St. Lawrence Bridge Company, Limited, *The Quebec Bridge Carrying the Transcontinental Line of the Canadian Government Railways over the St. Lawrence River near the City of Quebec* (n.p., n.d.), p. 14.
29. *The Quebec Bridge*, p. 14.

Chapter 4

30. "The New Sullivan Concentrator at Kimberly, B.C.," in *Investigations in Ore Dressing and Metallurgy, 1923* (Ottawa: Department of Mines, 1925), pp. 139–141 as quoted in Harold A. Innis, *Settlement and the Mining Frontier* (Toronto: Macmillan Company of Canada Limited, 1936), p. 303.
31. Various articles in *Pulp and Paper Magazine of Canada*, Special International Issue, February 1927.
32. John C. Van Nostrand, "The Queen Elizabeth Way: Public Utility Versus Public Space," *Urban History Review*, 12, 2 (October 1983): 1–23.

33. David Neufeld, "C.J. Mackenzie and the Challenge of the Prairies," in *Proceedings of the Canadian Society for Civil Engineering Annual Conference and the 7th Canadian Hydrotechnical Conference May 21–31, 1985*, Volume 2B, Structural, Construction, History, p. 490.

Chapter 5

34. J. De N. Kennedy, *History of the Department of Munitions and Supply Canada in the Second World War*, Volume 1, *Production Branches and Crown Companies* (Ottawa: King's Printer, 1950), p. 25.
35. Kennedy, *History of Munitions*, Vol.1, pp. 29–30.
36. C.J.S. Warrington and R.V.V. Nicholls, *A History of Chemistry in Canada* (Toronto: Chemical Institute of Canada and Sir Isaac Pitman and Sons (Canada) Limited, 1949), p. 157.
37. Sun Publishing Company Limited, *Industrial British Columbia 1945: Canada's Magnificent New Industrial Empire Geared to Efficient High Quality Production Ready for Immediate Post War Reconstruction* (Vancouver: Sun Publishing Company Limited, 1945).
38. *Industrial British Columbia*, p. 3.
39. Dominion Bridge Company Limited, *Of Tasks Accomplished: the Story of the Accomplishments of Dominion Bridge Company Limited and Its Wholly Owned Subsidiaries in World War II* (Montreal: Dominion Bridge Company Limited, 1945).

Chapter 7

40. Royal Commission on the Ocean Ranger Marine Disaster, Report One: *The Loss of the Submersible Drill Rig Ocean Ranger and Its Crew* (Ottawa: Supply and Services Canada, 1984), pp. iii–iv.
41. *Ocean Ranger*, p. 99.
42. Ibid.
43. Ibid.

Index

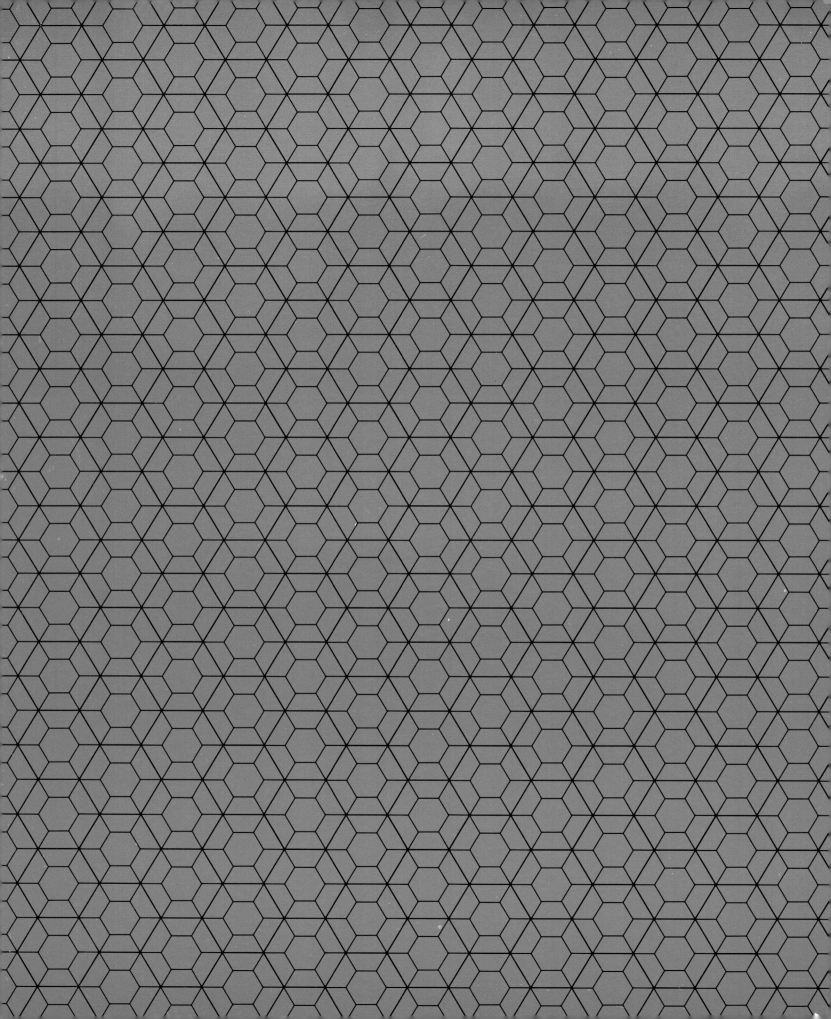